Answer Set Solving in Practice

Synthesis Lectures on Artificial Intelligence and Machine Learning

Editors
Ronald J. Brachman, *Yahoo! Labs*
William W. Cohen, *Carnegie Mellon University*
Peter Stone, *University of Texas at Austin*

Answer Set Solving in Practice
Martin Gebser, Roland Kaminski, Benjamin Kaufmann, and Torsten Schaub
2012

Planning with Markov Decision Processes: An AI Perspective
Mausam and Andrey Kolobov
2012

Active Learning
Burr Settles
2012

Computational Aspects of Cooperative Game Theory
Georgios Chalkiadakis, Edith Elkind, and Michael Wooldridge
2011

Representations and Techniques for 3D Object Recognition and Scene Interpretation
Derek Hoiem and Silvio Savarese
2011

A Short Introduction to Preferences: Between Artificial Intelligence and Social Choice
Francesca Rossi, Kristen Brent Venable, and Toby Walsh
2011

Human Computation
Edith Law and Luis von Ahn
2011

Answer Set Solving in Practice

Martin Gebser, Roland Kaminski, Benjamin Kaufmann, and Torsten Schaub

ISBN: 978-3-031-00433-9   paperback
ISBN: 978-3-031-01561-8   ebook

DOI 10.1007/978-3-031-01561-8

A Publication in the Springer series
SYNTHESIS LECTURES ON ARTIFICIAL INTELLIGENCE AND MACHINE LEARNING

Lecture #19
Series Editors: Ronald J. Brachman, Yahoo! Labs
              William W. Cohen, Carnegie Mellon University
              Peter Stone, University of Texas at Austin
Series ISSN
Synthesis Lectures on Artificial Intelligence and Machine Learning
Print 1939-4608   Electronic 1939-4616

Answer Set Solving in Practice

Martin Gebser, Roland Kaminski, Benjamin Kaufmann, and Torsten Schaub
University of Potsdam

SYNTHESIS LECTURES ON ARTIFICIAL INTELLIGENCE AND MACHINE LEARNING #19

ABSTRACT

Answer set programming (ASP) is a declarative problem solving approach, initially tailored to modeling problems in the area of knowledge representation and reasoning (KRR). More recently, its attractive combination of a rich yet simple modeling language with high-performance solving capacities has sparked interest in many other areas even beyond KRR.

This book presents a practical introduction to ASP, aiming at using ASP languages and systems for solving application problems. Starting from the essential formal foundations, it introduces ASP's solving technology, modeling language and methodology, while illustrating the overall solving process by practical examples.

KEYWORDS

answer set programming, declarative problem solving, logic programming

To *Pascal*, and all *the Ones* who enriched our lives in a lasting way.

Contents

List of Figures

List of Tables

List of Listings

List of Algorithms

Preface

The roots of this book lie in a series of lectures on answer set programming (ASP) started in 2001 at the University of Potsdam. Back then, ASP was still in its infancy and its future was far from clear. With ASP, the area of nonmonotonic reasoning was melting into logic programming, and the relationship to Boolean constraint solving, in particular, satisfiability testing was just about to be discovered. ASP's driving force was the ASP solver *smodels* along with its grounder *lparse* that provided effective means to experiment with ASP and to bridge the gap between theory and practice. And even more systems were emerging at the time. This spirit made us launch a course on ASP in order to leave behind glassy students' eyes worrying whether they should bother about the flying capacities of Tweety the penguin. And indeed, there appeared this spark, nourished by the mystery of "automated problem solving," that drew the interest of students with various backgrounds. In some years, our ASP lectures had more students than our AI class!

Although this book was put together by the last author, it reflects the joint effort of all authors conducted over the last years. Hence, he is the one to blame for bad writing! In fact, the course leading up to this book would have never taken off without the groundbreaking tutorials of Vladimir Lifschitz, Ilkka Niemelä, and Wolfgang Faber at the time. Of course, we are indebted to our colleagues and co-authors who accompanied our research over the last decade, Alessandra Mileo, Andreas Schwill, André Flöter, André Neumann, Anne Siegel, Arne König, Benjamin Andres, Benjamin Lüpfert, Bettina Schnor, Bob Mercer, Carito Guziolowski, Christian Anger, Christian Drescher, Christian Haubelt, Christophe Bobda, Enrico Ellguth, Farid Benhammadi, Gerhard Brewka, Hans Tompits, Holger Jost, Jacques Nicolas, Javier Romero, Jean Gressmann, Jim Delgrande, Joachim Selbig, Joohyung Lee, Jörg Pührer, Kathrin Konczak, Katsumi Inoue, Kewen Wang, Lars Schneidenbach, Marina De Vos, Marius Schneider, Martin Brain, Maurice Pagnucco, Max Ostrowski, Miroslaw Truszczynski, Mona Gharib, Murat Knecht, Oliver Matheis, Orkunt Sabuncu, Pascal Nicolas, Paul Borchert, Peter-Uwe Zettiér, Philipp Obermeier, Philippe Besnard, Philippe Veber, Richard Tichy, Roberto Bisiani, Sabine Hübner, Stefan Brüning, Stefan Woltran, Stefan Ziller, Steve Dworschak, Sven Thiele, Thibaut Henin, Thomas Krennwallner, Thomas Leupold, Thomas Linke, Tobias Schubert, Tomi Janhunen, Torsten Grote, Vladimir Sarsakov, Wolfgang Severin, Yan Zhang, Yuliya Lierler, Yves Moinard, and presumably many more, who slipped our minds. Also, we are truly grateful to the Deutsche Forschungsgemeinschaft (DFG) for supporting our basic research over the last decade.

Finally, it goes without saying that our work would have been impossible without the constant support of those close to us, our friends, families, and beloved ones.

Martin Gebser, Roland Kaminski, Benjamin Kaufmann, and Torsten Schaub
November 2012

About this book

The goal of this book is to enable people to use answer set programming (ASP) for problem solving while understanding the functioning of the underlying solving machinery. To this end, we focus on modeling and solving techniques and restrict ourselves to giving essential theoretical foundations. In fact, the distinguishing feature of ASP is its attractive combination of a rich yet simple modeling language with high-performance solving capacities stemming from the area of Boolean constraint solving. We take up the latter to provide a uniform view on both constituents and to facilitate insights into their interaction. For illustrating ASP's modeling and solving capacities, we take advantage of the suite of ASP systems provided at `potassco.sourceforge.net`. This also guides our formal development, which is driven by the elaboration of the underlying logical and algorithmic principles.

The outline of the book is as follows.

Chapter 1 starts by motivating the approach of answer set programming (ASP) and by contrasting it to traditional logic programming and satisfiability testing. Moreover, it provides a quick start to the ASP systems at `potassco.sourceforge.net` for the reader wanting to practically experience ASP on the go.

Chapter 2 lays the essential foundations for the remaining chapters. We begin with a brief introduction to our terminology and the basic definitions underlying ASP. The principal part is dedicated to the syntax and semantics of ASP's modeling language. The remainder deals with computational issues and derives a first algorithm from ASP's basic definitions.

Chapter 3 takes up the modeling language presented in the previous chapter and introduces ASP's modeling methodology along with its solving process by carefully developing solutions to some well-known examples.

Chapter 4 describes the basic functioning of the first part of ASP's solving process, the systematic instantiation of first-order variables, referred to as grounding. We sketch the underlying algorithms and illustrate the obtained expressive power by encoding a universal Turing machine and presenting meta programming techniques in ASP.

Chapter 5 lays the theoretical foundations for understanding the multitude of inferences involved in ASP solving. We consider in turn alternative characterizations of stable models. Starting from an axiomatic viewpoint, we successively make the underlying inferences more and more precise, ultimately distilling a formal characterization capturing the inferences drawn by the solving algorithms presented in the next chapter.

Chapter 6 builds upon the constraint-based characterization of stable models developed in the last chapter for developing an algorithmic framework relying on modern Boolean constraint technology involving conflict-driven learning.

Chapter 7 describes the grounder *gringo* and the solver *clasp*. Both systems serve not only as illustrative models but also provide us with insights into the architecture and functioning of advanced ASP technology. The chapter is complemented by a survey of further ASP systems furnished by the suite of ASP systems at potassco.sourceforge.net that build upon *gringo* and *clasp*.

Chapter 8 continues the introduction to ASP modeling in Chapter 3 and elaborates upon advanced modeling techniques in view of the grounding, solving, and system expertise built up in the intermediate chapters. We identify several modeling patterns and discuss their impact on the ASP solving process. Although such knowledge is not needed for basic ASP modeling, it is highly beneficial in view of scalability.

Chapter 9 concludes this book by summarizing its key aspects and designating some future research challenges.

Each chapter closes with a section surveying the respective bibliography and indicating further reading.

The chapters can roughly be classified into three categories: Fundamental Chapters 2 and 5, Chapters 3 and 8 on modeling in ASP, and Chapters 4, 6, and 7 on ASP's solving process. Although the material in this book is laid out for sequential reading, other paths are possible. Readers specifically

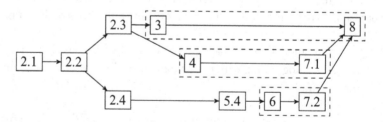

Figure 1: Selected dependencies for reading.

interested in grounding or solving, respectively, can follow the dependencies in Figure 1 leading from Section 2.1 to Chapter 4/Section 7.1 for grounding or Chapter 6/Section 7.2 for solving (indicated by dashed boxes). Readers interested in modeling need little prerequisites to get a good acquaintance with the basic modeling methodology of ASP presented in Chapter 3. However, the study of the advanced modeling techniques in Chapter 8 benefits from some familiarity with the ASP solving process in order to understand their interaction.

The material presented in this book is accompanied by a package of slides. Both have been used in a hebdomadary lecture of two hours per week. This and more material is available at `potassco.sourceforge.net/teaching.html`.

Martin Gebser, Roland Kaminski, Benjamin Kaufmann, and Torsten Schaub
November 2012

CHAPTER 1

Motivation

Answer Set Programming is an approach to *declarative problem solving*. Rather than solving a problem by telling a computer *how to solve the problem*, the idea is simply to describe *what the problem is* and leave its solution to the computer. This approach is illustrated in Figure 1.1.[1] Starting from

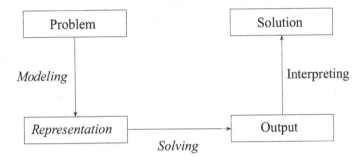

Figure 1.1: Declarative problem solving.

the original problem, traditional programming results in an executable computer program. On the contrary, *modeling* aims at creating a formal *representation* of the original problem. Instead of merely executing a resulting computer program, the obtained problem representation is used in *solving* to extract an implicit state space that is explored by sophisticated search algorithms for finding a solution to the original problem.

    The basic idea of *Answer Set Programming* (ASP) is to express a problem in a logical format so that the models of its representation provide the solutions to the original problem. The resulting models are referred to as *answer sets*. The actual notion of a model is determined by the logic of choice. Although this broad view attributes to ASP the character of a general logical constraint processing paradigm, the term ASP is nowadays mainly associated with theories in the syntax of *logic programs* under the *stable models semantics* as introduced by Michael Gelfond and Vladimir Lifschitz in 1988. While such programs resemble Prolog programs, they are however treated by quite different computational mechanisms. Indeed ASP can be regarded as a much better fit to the original motivation of logic programming by strictly separating logic from control.

    Comparing ASP to a traditional logic programming language such as Prolog reveals some key differences. Prolog is based on top-down query evaluation in the tradition of Automated Theorem

---

[1]Entities diverging from traditional programming are given in *italic*.

Proving. Variables are dealt with via unification and (nested) terms are used as basic data structures. A solution is usually extracted from the instantiation of the variables in a successful query. As mentioned, solutions are captured by models in ASP, and instead computed in a bottom-up fashion. Variables are systematically replaced by using database techniques. Hence tuples and (flat) terms are the preferred data structures. More generally, Prolog constitutes a full-fledged programming language and thus equips a user with control over program execution, whereas ASP fully decouples a problem's specification from how its solution is found.

Even though the formal roots of ASP indeed lie in Logic Programming, it was tailored right from the beginning to problem solving in the field of Knowledge Representation and Reasoning. The accompanying desire for transparent and elaboration-tolerant representation languages along with the significant advance in Boolean Constraint Solving were then the two major impetuses to ASP's distinguished combination of a rich yet simple modeling language with high-performance solving capacities.

The effectiveness of modern ASP solvers would have been impossible without the great progress in Boolean Constraint Solving, mainly conducted in the area of propositional Satisfiability Testing (SAT). As well, the breakthrough in model-oriented problem solving was pioneered in the context of SAT: Kautz and Selman represented planning problems in 1992 as propositional theories so that models (not proofs) described solutions and demonstrated that this approach was competitive with state-of-the-art planning systems at the time. Logically, the difference between ASP and SAT boils down to the logic of choice and its associated notion of modelhood. Informally, stable models can be regarded as distinguished (classical) models of a theory, in which each true atom must be provable. This constructive flavor of ASP translates into more succinct problem representations than available in SAT. From a representational viewpoint, this semantic difference reduces to *closed world reasoning*, that is, considering propositions as false unless proven otherwise. From the perspective of computational complexity, both ASP and SAT allow for expressing search problems in *NP*. The disjunctive extension of ASP also captures problems in NP^{NP}. System-wise the focus of SAT lies in solving, while ASP is moreover concerned with modeling. As a consequence, ASP solving comprises an initial grounding phase in which first-order problem representations are translated into a propositional format. This propositionalization is accomplished by highly efficient grounders based on Database technology.

Putting things together, the overall ASP solving process can be summarized as in Figure 1.2. A problem is modeled in the syntax of (first-order) logic programs. Then, ASP solving proceeds in two steps. First, a grounder generates a finite propositional representation of the input program. After that, a solver computes the stable models of the propositional program. Finally, the solution is read off the resulting stable models. Let us illustrate this process by means of the simplistic yet authentic program in Listing 1.1.

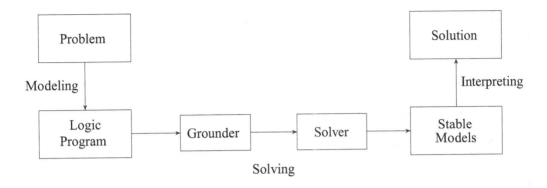

Figure 1.2: ASP solving process.

Listing 1.1: The problem of traveling out of Berlin in ASP (roads.lp)

```
1   road(berlin,potsdam).
2   road(potsdam,werder).
3   road(werder,brandenburg).
4   road(X,Y) :- road(Y,X).

6   blocked(werder,brandenburg).

8   route(X,Y) :- road(X,Y), not blocked(X,Y).
9   route(X,Y) :- route(X,Z), route(Z,Y).

11  drive(X) :- route(berlin,X).

13  #hide.
14  #show drive/1.
```

A logic program consists of *facts* (as in Lines 1–3 and 6) and *rules* (as in Lines 4, 8, 9, and 11), each of which is terminated by a period '.'. The connectives ':-' and ',' can be read as *if* and *and*, respectively. A statement commencing with 'not' is satisfied unless its enclosed proposition is found to be true. Lines 13 and 14 are directives to the grounder and solver, respectively, and thus do not belong to the actual program. In our example, the facts in Lines 1–3 are meant to provide road information among cities, represented by constants berlin, brandenburg, potsdam, and werder. Line 4 tells us that each road can be taken in both directions. For this purpose, we use variables denoted by uppercase letters X and Y that range over the aforementioned constant symbols. Line 6 indicates an obstruction between Werder and Brandenburg. According to the rule in Line 8, roads are routes unless they are blocked. Alternatively, routes can be composed from other routes (Line 9). The predicate drive tells us which cities are reachable from Berlin. Finally, Lines 13 and 14 direct the solver to project models onto the satisfied instances of predicate `drive`. In Listing 1.2,[2] it is shown how the logic program `roads.lp` is processed by the ASP grounder *gringo* and passed to the ASP solver *clasp*. (The argument 0 makes the solver compute all stable models.)

Listing 1.2: The solution of traveling out of Berlin in ASP

```
$ gringo roads.lp | clasp 0
clasp version 2.0.5
Reading from stdin
Solving...
Answer: 1
drive(berlin) drive(werder) drive(potsdam)
SATISFIABLE

Models     : 1
Time       : 0.000s (Solving: 0.00s 1st Model: 0.00s Unsat: 0.00s)
CPU Time   : 0.010s
```

Looking at the obtained answer, we observe that only Berlin, Potsdam, and Werder are reachable destinations (due to the blocked road between Werder and Brandenburg).

1.1   QUICKSTART

The concepts introduced in this book constitute the major ingredients of the ASP systems gathered at `potassco.sourceforge.net`. *Hence before taking the reader on a guided tour through our conceptual toolbox, we wish to relieve her impatience a bit and provide a sneak preview into the usage of our systems. This should also furnish her some basic experimental skills that allow her to explore things on the go. Enjoy!*

As an introductory example, we consider a simple Towers of Hanoi puzzle consisting of three pegs and four disks of different size. As shown in Figure 1.3, the goal is to move all disks from the left peg to the right one. Only the topmost disk of a peg can be moved at a time. Furthermore, a disk

Figure 1.3: Towers of Hanoi: initial and goal situation.

[2]The symbol $ stands for a computer terminal's command prompt.

cannot be moved to a peg already containing a disk of smaller size. Although there is an efficient algorithm to solve our simple puzzle, we do not exploit it and below merely specify conditions for sequences of moves being solutions.

Problem representation Following good practice in ASP, we separately provide an instance and an encoding (applying to every instance) of the following problem: given an initial placement of the disks, a goal situation, and a number n, decide whether there is a sequence of n moves that achieves the goal. We show that this problem can be elegantly specified in ASP and solved by domain-independent tools like *gringo* and *clasp*.

Problem instance We describe the pegs and disks of a Towers of Hanoi puzzle via facts over the unary predicates peg/1 and disk/1.[3] Disks are numbered by consecutive integers starting at 1, where a disk with a smaller number is considered to be larger than a disk with a greater number. The names of the pegs can be arbitrary; in our case, we use a, b, and c. Furthermore, the predicates init_on/2 and goal_on/2 describe the initial and the goal situation, respectively. Their arguments, the number of a disk and the name of a peg, determine the location of a disk in the respective situation. Finally, the constant moves/0 specifies the number of moves in which the goal must be achieved. The Towers of Hanoi puzzle shown in Figure 1.3 can be described by the facts in Listing 1.3.

Listing 1.3: Towers of Hanoi problem instance (tohI.lp)

```
1  #const moves=15.

3  peg(a;b;c).
4  disk(1..4).
5  init_on(1..4,a).
6  goal_on(1..4,c).
```

The very first line is an extra-logical directive to the grounder,[4] providing default value 15 for constant moves. Note that the ; in the Line 3 is syntactic sugar expanding the statement into three facts: peg(a), peg(b), and peg(c). Similarly, 1..4 used in Lines 4–6 refers to an interval abbreviating distinct facts over the four values: 1, 2, 3, and 4. Observe that the initial and the goal situation are specified by associating disks with pegs, while relative locations of disks are given implicitly: among disks sharing a peg, the disk with the greatest number rests on top of the one with the second greatest number, provided the latter exists, and so on. In summary, the given facts describe the Towers of Hanoi puzzle in Figure 1.3 along with the requirement that the goal ought to be achieved with 15 moves.

Problem encoding We now proceed by encoding Towers of Hanoi via rules containing variables (whose names start with uppercase letters). Such an encoding is independent of a particular

---

[3]We use p/n to indicate that predicate p has arity n.
[4]Alternatively, such definitions can be supplied (or overridden) via the command line option --const of *gringo*.

instance. Typically, an encoding can be logically partitioned into "generating," "defining," and "testing" parts (see Section 3.2). An additional "displaying" part allows for projecting the output to atoms characterizing a solution, thereby suppressing auxiliary predicates. We follow this methodology and mark the respective parts via comment lines beginning with % in Listing 1.4.

Listing 1.4: Towers of Hanoi problem encoding (tohE.lp)

```
1   %    generating part
2   1 { move(D,P,T) : disk(D) : peg(P) } 1 :- T = 1..moves.

4   %    defining part
5   move(D,T)              :- move(D,_,T).
6   on(D,P,0)              :- init_on(D,P).
7   on(D,P,T)              :- move(D,P,T).
8   on(D,P,T+1)            :- on(D,P,T), not move(D,T+1), T < moves.
9   blocked(D-1,P,T+1)     :- on(D,P,T), T < moves.
10  blocked(D-1,P,T)       :- blocked(D,P,T), disk(D).

12  %    testing part
13  :- move(D,P,T), blocked(D-1,P,T).
14  :- move(D,T), on(D,P,T-1), blocked(D,P,T).
15  :- not 1 { on(D,P,T) } 1, disk(D), T = 1..moves.

17  :- goal_on(D,P), not on(D,P,moves).

19  %    displaying part
20  #hide.
21  #show move/3.
```

Note that the variables D, P, and T are used to refer to disks, pegs, and the number of a move, respectively. While all occurrences of these variables refer within each rule to the same variable, the token _ (not followed by any letter) stands for an *anonymous variable* that does not recur anywhere. (This is as if a new variable name is invented on each occurrence of _.) In contrast, moves is a constant indicating the length of the sequence of moves.

The generating part, describing solution candidates, consists of the rule in Line 2. It expresses that exactly one move of a disk D to some peg P must be executed at each point T in time (other than 0). The head of the rule (left of :-) is a so-called cardinality constraint (detailed in Section 2.3). It consists of a set of literals, expanded using the predicates behind the colons (detailed in Section 2.3), along with a lower and an upper bound. The cardinality constraint is satisfied if the number of true elements is between the lower and the upper bound (both inclusive). In our case, both bounds are 1. Since the cardinality constraint occurs as the head of a rule, it allows for deriving ("guessing") atoms over the predicate move/3. In the body (right of :-), we use the assignment 'T = 1..moves' to refer to each time point T from 1 to the maximum, given by moves. We have thus characterized all sequences of moves as solution candidates for Towers of Hanoi. Up to now, we have not yet imposed any further conditions, for instance, that a larger disk must not be moved on top of a smaller one.

The defining part in Lines 5–10 contains rules deriving auxiliary predicates providing properties of a solution candidate at hand. (Such properties are investigated in the testing part described below.) The rule in Line 5 simply projects moves to disks and time points. The resulting predicate move/2 can be used whenever the target peg is irrelevant. In this way, one of its atoms subsumes three possible cases. Furthermore, the predicate on/3 captures the state of a Towers of Hanoi puzzle at each time point. To this end, the rule in Line 6 identifies the locations of disks at time point 0 by inspecting the initial state. State transitions are modeled by the rules in Lines 7 and 8. While the former specifies the direct effect of a move at time point T, that is, the considered disk D is relocated to the target peg P, the latter describes inertia: the location of a disk D carries forward from time point T to T+1 if D is not moved at T+1. Observe that 'T < moves' in Line 8 prevents the derivation of disk locations beyond the maximum time point.

Finally, we employ some more sophisticated concepts to facilitate the tests in Lines 13 and 14. For this purpose, we define the auxiliary predicate blocked/3 to indicate that a smaller disk (with a number greater than D-1) is located on a peg P. The rule in Line 9 derives this condition for time point T+1 from on(D,P,T), provided that T is not the maximum time point. The rule in Line 10 further propagates the status of being blocked along larger disks.[5] Note that we also mark 'D-1 = 0', not referring to any disk, as blocked, which is convenient for eliminating redundant moves in the testing part described below. To illustrate this approach, let us inspect the initial situation displayed in Figure 1.3. Given the facts over init_on/2 in Listing 1.3, the rule in Line 6 yields on(1,a,0), on(2,a,0), on(3,a,0), and on(4,a,0). Provided that '0 < moves', we further derive blocked(0,a,1), blocked(1,a,1), blocked(2,a,1), and blocked(3,a,1) via the rule in Line 9 (and 10). This tells us that, apart from disk 4, any location on peg a is occupied and should thus not be used in moves at time point 1. In particular, note that the movable disk 4 ought not be put back to peg a because its adjacent location 3 is blocked there (due to disk 4 itself).

The testing part consists of the integrity constraints in Lines 13–15 and 17. Such rules have no head atom. Logically, such an empty head can be understood as a contradiction; hence, the body atoms cannot all be satisfied simultaneously. Integrity constraints serve as rules of denial eliminating unintended solution candidates. The first integrity constraint in Line 13 asserts that a disk D must not be moved to a peg P, if D-1 is blocked at time point T. This excludes moves putting a larger disk on top of a smaller one (whose number is greater). And since the adjacent location D-1 is blocked on the former peg of D (in view of the rule in Line 9), it also prevents putting a disk back on its previous location. Similarly, the integrity constraint in Line 14 expresses that a disk D cannot be moved at time point T, if it is blocked by some smaller disk on the same peg P. Note that we use move(D,T) here because the target of an illegal move does not matter in this context. The integrity constraint in Line 15 asserts that, for every disk D and time point T, there is exactly one peg P such

---

[5]The interested reader may observe that the extension of blocked/3 is linear in the number of disks, while an encoding relying on pairwisely comparing positions would lead to a quadratic behavior.

that on(D,P,T) holds. Although this condition is implied by the definition of on/3 in Lines 7 and 8 with respect to the moves in a solution, making such knowledge explicit via an integrity constraint turns out to improve the solving efficiency. Finally, the integrity constraint in Line 17 addresses the goal situation that must be achieved at maximum time point moves.

Lastly, the two directives of the displaying part in Lines 20 and 21 indicate that only atoms over the predicate move/3 are to be printed. It suppresses all predicates used to describe problem instances as well as atoms stemming from auxiliary predicates move/2, on/3, and blocked/3. This is for more convenient reading of a solution, given that it is fully determined by atoms over move/3 (see also Section 7.1).

Problem solving   We are now ready to solve our Towers of Hanoi puzzle. To compute a stable model representing a solution, we may invoke the following command. [6]

```
$ gringo tohI.lp tohE.lp | clasp
```

The filenames tohI.lp and tohE.lp refer to the programs in Listings 1.3 and 1.4, respectively.

An alternative to this is offered by *clingo*, combining *clasp* and *gringo* in a monolithic system.

```
$ clingo tohI.lp tohE.lp
```

The output of feeding the result of the grounder *gringo* into the solver *clasp* looks (somewhat) as shown in Listing 1.5.

Listing 1.5: Solving the Towers of Hanoi problem

```
$ gringo tohI.lp tohE.lp | clasp
clasp version 2.0.5
Reading from stdin
Solving...
Answer: 1
move(4,b, 1) move(3,c, 2) move(4,c, 3) move(2,b, 4) move(4,a, 5) \
move(3,b, 6) move(4,b, 7) move(1,c, 8) move(4,c, 9) move(3,a,10) \
move(4,a,11) move(2,c,12) move(4,b,13) move(3,c,14) move(4,c,15)
SATISFIABLE

Models    : 1+
Time      : 0.009s (Solving: 0.00s 1st Model: 0.00s Unsat: 0.00s)
CPU Time  : 0.000s
```

The first line provides version information concerning the ASP solver *clasp*. No such information is given by the grounder *gringo* because its output is read by the solver. Similarly, the next two lines provide status information of *clasp*. Once "solving" has finished, 'Answer: 1' indicates that

---

[6]All systems, *clasp*, *gringo*, and *clingo* are freely available at potassco.sourceforge.net.

the (output) atoms of the first stable model follow in the line below. We use here the symbol \ to indicate that all atoms over move/3 actually belong to a single line. Note that the order in which atoms are printed does not have any meaning (and the same applies to the order in which stable models are found). Here, we show the instances of move/3 in the order of time points, so that we can easily read off the following solution: first move disk 4 to peg b, second move disk 3 to peg c, third move disk 4 to peg c, and so on. Below this solution, the solver reports the problem's satisfiability status, viz. SATISFIABLE[7] in our example. The 1+ in the line starting with Models tells us that one stable model has been found.[8] The two final lines report various timing statistics, distinguishing wall clock and CPU time. More information about available options, for instance, how to get extended statistics output, can be obtained via option --help. See also Section 7.2 on how to interpret *clasp*'s extended statistics.

Recapitulation   For solving our Towers of Hanoi puzzle, we first provided facts representing an instance. Although we did not discuss the choice of predicates, an appropriate instance representation is already part of the modeling in ASP and not always as straightforward as here.[9] Second, we provided an encoding of the problem applying to any instance. The encoding consisted of parts generating solution candidates, deriving their essential properties, testing that no solution condition is violated, and finally projecting the output to characteristic atoms. With the encoding at hand, we could use off-the-shelf ASP tools to solve our instance, and the encoding can be reused for any further instance that may arise in the future.

1.2   REFERENCES AND FURTHER READING

The first monograph on ASP was written by Baral in (2003); further comprehensive accounts can be found in Eiter et al. (2009), Ferraris and Lifschitz (2005), Gelfond (2008), Lifschitz (1996, 2004). The handbooks of Knowledge Representation and Reasoning (Lifschitz et al., 2008), Satisfiability Testing (Biere et al., 2009), and Constraint Programming (Rossi et al., 2006), respectively, provide rich sources on ASP-related technology. The same applies to the standard literature on Logic Programming (Lloyd, 1987) and Database Systems (Ullman, 1988).

The term *Answer Set Programming* was coined in the late nineties in Lifschitz (1999), Marek and Truszczyński (1999), Niemelä (1999). ASP is sometimes also referred to as AnsProlog or simply A-Prolog (Baral, 2003, Gelfond, 2002, Gelfond and Leone, 2002). A recent account of ASP is given by Brewka et al. (2011).

The concept of *elaboration tolerance* was advocated by John McCarthy in (1998); it is defined as "*the ability to accept changes to a person's or a computer program's representation of facts about a subject without having to start all over.*"

---

[7]Other possibilities include UNSATISFIABLE and UNKNOWN, the latter in case of an abort.

[8]The + indicates that the solver has not exhaustively explored the search space (but stopped upon finding a stable model), so that further stable models may exist.

[9]When trying to solve a problem and finding it cumbersome to specify appropriate rules, it may be worthwhile to recheck the instance format and alter it to something more appropriate.

The field of Logic Programming was initiated by the work of Alain Colmerauer and Robert Kowalski in the mid-seventies. The separation of logic and control in Logic Programming was formulated by the slogan *"Algorithm = Logic + Control"* (Kowalski, 1979). Although both Prolog and ASP strive for declarativeness, Prolog aims at being a full-fledged programming language and thus allows the programmer to influence the program's execution via the order of its rules and their body literals.

Historically, there is also a strong connection between Logic Programming and Deductive Databases, as witnessed by Minker (1988).

The representational edge of ASP over SAT is due to its more stringent semantics. Niemelä showed in (1999) that SAT can be translated modularly into ASP but not vice versa. Lifschitz and Razborov (2006) proved (subject to complexity-theoretic assumptions) that every vocabulary-preserving translation from ASP to SAT must be exponential (in the worst case) but not vice versa. For a thorough analysis in terms of complexity theory, we refer the reader to Dantsin et al. (2001), Schlipf (1995).

Closed world reasoning is a salient feature of commonsense reasoning; it is fundamental to Database Systems (Reiter, 1978) and Logic Programming (Lloyd, 1987), where it is at the heart of *negation-as-failure*. A broad exploration of the phenomenon along with its various formal accounts is explored in depth in the area of Nonmonotonic Reasoning (Besnard, 1989, Ginsberg, 1987, Marek and Truszczyński, 1993).

The interpretation of closed world reasoning in terms of (inductive) definitions is worked out in Denecker and Ternovska (2008) and leads to a solving paradigm closely related to ASP (Mariën et al., 2004). Interesting combinations of modeling and solving in SAT are provided by *kodkod* (Torlak and Jackson, 2007), a constraint solver for first-order logic with relations, and *npspec* (Cadoli and Schaerf, 2005), a Datalog-like language based on minimal models semantics.

Although the *stable models semantics* was originally proposed as a semantics for logic programs (Gelfond and Lifschitz, 1988, 1991), it was strongly influenced by systems for Nonmonotonic Reasoning, in particular, Autoepistemic (Moore, 1985) and Default Logic (Reiter, 1980). (See also Section 5.5.) The equivalent *default models semantics* was independently proposed by Bidoit and Froidevaux in (1987). While the intuition of the former stemmed from Autoepistemic Logic (see also Gelfond (1987)), the latter semantics was derived from Default Logic. The relationship to Autoepistemic Logic is established in Gelfond and Lifschitz (1988). Also, Gelfond and Lifschitz show in (1991) that logic programs under stable models semantics constitute a fragment of Default Logic. A similar reduction was given in Bidoit and Froidevaux (1987).

The light constructive nature of the stable models semantics was established by Pearce in (1996), leading to the encompassing framework of Equilibrium Logic (Pearce, 2006). The latter is based on the strongest super-intuitionistic logic that is properly contained in classical logic, called the logic of *Here-and-There* (Gödel, 1932, Heyting, 1930).

CHAPTER 2

Introduction

This chapter gives the formal foundations for the material presented in this book. We lay out salient concepts and fix corresponding notations. We start by providing some terminology for essential logical concepts in Section 2.1; less broadly used concepts are introduced where needed. Section 2.2 provides a compact introduction to the basic propositional syntax and semantics of ASP. We then successively extend the core syntax with several language constructs in view of improving the ease of modeling. Similarly, we build upon the core semantics in Section 2.4 for developing a first scheme for computing the stable models of a logic program. The final section provides some historical remarks and gives references for further reading.

2.1 LOGICAL PRELIMINARIES AND TERMINOLOGY

We presuppose some acquaintance with the basics of Logic and Logic Programming. We thus restrict ourselves to the introduction of relevant notation and refer the interested reader for introductory material to the literature (see Section 2.5).

Although we use standard notation to refer to logical concepts, we adapt it to the various levels of description in order to stress the respective view (see Table 2.1).

Table 2.1: Notational conventions according to level of description							
	true, false	if	and	or	iff	default negation	classical negation
source code		:-	,	|		not	−
logic program		←	,	;		∼	¬
formula	⊤, ⊥	→	∧	∨	↔	∼	¬

We consider languages having a signature composed of symbols for predicates, functions, and variables. We usually denote

- predicate symbols by lowercase letters, such as $a, b, \ldots$ for predicates of zero arity (also called propositions), and otherwise use $p, q, \ldots$ or strings starting with a lowercase letter, such as *hot* or *hasCookie*,

- function symbols by lowercase letters, such as $c, d, \ldots$ for constants, and otherwise $f, g, \ldots$, or *size*, and

- variable symbols by uppercase letters, such as X, Y, Z, or strings starting with an uppercase letter, like *Mother*.

Each predicate and function symbol has an associated arity, n, which is sometimes made explicit by writing p/n or f/n, respectively.

Terms and atoms are defined in the usual way. Variable-free terms and atoms are said to be *ground*. A ground instance of an atom is obtained by replacing all of its variables by ground terms. We often identify ground atoms with propositions and also denote them by lowercase letters, like $a, b, \ldots$.

We let T and F stand for the Boolean truth values. A (total) two-valued *interpretation*, mapping all ground atoms to T and F, is represented by the set of its true atoms. For instance, the interpretation $\{p(1) \mapsto T, p(2) \mapsto F, a \mapsto T\}$, is represented by $\{p(1), a\}$. This allows us to consider an interpretation to be smaller than another, if one is a subset of another.

For capturing partial interpretations, we use three-valued interpretations and represent them as pairs, consisting of all true and false ground atoms, while leaving the remaining undefined ones implicit. For example, the three-valued interpretation $(\{p(1)\}, \{a\})$ assigns T to $p(1)$ and F to a, respectively, and regards all other ground atoms as undefined. A three-valued interpretation is total (and two-valued), if it assigns either T or F to all ground atoms; otherwise, it is said to be partial.

For addressing computational issues, we rely on Boolean assignments. An ordered *assignment* A over a domain, $dom(A)$, is a sequence $(\sigma_1, \ldots, \sigma_n)$ of entries σ_i of the form Tv_i or Fv_i, where $v_i \in dom(A)$ for $1 \leq i \leq n$. An assignment's domain consists of logical entities, usually going beyond the set of atoms. We often refer to them as propositional variables. An entry Tv or Fv expresses that v is assigned T or F, respectively. We denote the complement of an entry σ by $\overline{\sigma}$, that is, $\overline{Tv} = Fv$ and $\overline{Fv} = Tv$. Analogously, an unordered assignment is a set (rather than a sequence) of entries. We sometimes abuse notation and identify an ordered assignment with its unordered counterpart given by the set of its contained entries. Given this, we access the true and the false (propositional) variables in A via $A^T = \{v \in dom(A) \mid Tv \in A\}$ and $A^F = \{v \in dom(A) \mid Fv \in A\}$. We say that A is contradictory if $A^T \cap A^F \neq \emptyset$; otherwise, A is non-contradictory. Furthermore, A is total if it is non-contradictory and $A^T \cup A^F = dom(A)$. Note that an unordered assignment A over atoms can be regarded as a partial interpretation (T, F) where $Ta \in A$ iff $a \in T$ and $Fa \in A$ iff $a \in F$ for all ground atoms a.

A minimal element of a partially ordered set is an element that is not greater than any other element; maximal elements are defined analogously. We often consider power sets ordered by set inclusion, and refer to the minimal elements as being $\subseteq$-minimal.

Finally, we often make use of graph-theoretic concepts. A (directed) *graph* is an ordered pair (V, E) comprising a set V of vertices together with a set $E \subseteq V \times V$ of edges. A path in a graph is a sequence of vertices such that there is an edge for each succeeding pair of vertices in the sequence. A directed graph is strongly connected, if all its vertices are pairwisely connected by some path. The strongly connected components of a directed graph are its maximal strongly connected subgraphs. Such a component is non-trivial, if it contains some edge.

Further (logical) concepts are introduced where needed.

2.2   BASIC SYNTAX AND SEMANTICS

We give a compact formal introduction to propositional logic programs under stable models semantics. Most concepts are developed in more detail in the remainder of the book.

A propositional normal *logic program* over a set $\mathcal{A}$ of ground atoms is a finite set of normal *rules* of the form

$$a_0 \leftarrow a_1, \ldots, a_m, {\sim}a_{m+1}, \ldots, {\sim}a_n \tag{2.1}$$

where $0 \leq m \leq n$ and each $a_i \in \mathcal{A}$ is a ground atom for $0 \leq i \leq n$. A *literal* is an atom a or its *default negation*[1] ${\sim}a$.

As a first simple example, consider Logic program P_1.

$$P_1 \;=\; \left\{ \begin{array}{l} a \leftarrow \\ c \leftarrow {\sim}b, {\sim}d \\ d \leftarrow a, {\sim}c \end{array} \right\}$$

For a rule r as in (2.1), let

$$
\begin{aligned}
head(r) &= a_0 && \text{be the } head \text{ of } r, \text{ and} \\
body(r) &= \{a_1, \ldots, a_m, {\sim}a_{m+1}, \ldots, {\sim}a_n\} && \text{be the } body \text{ of } r.
\end{aligned}
$$

The intuitive reading of r is that $head(r)$ must be true if $body(r)$ holds, that is, if $a_1, \ldots, a_m$ are (provably) true and if $a_{m+1}, \ldots, a_n$ are (possibly) false. If $body(r) = \emptyset$, r is called a *fact*, and we often omit '$\leftarrow$' when writing facts.

Given a set X of literals, let $X^+ = \{p \in \mathcal{A} \mid p \in X\}$ and $X^- = \{a \in \mathcal{A} \mid {\sim}a \in X\}$. For $body(r)$, we then have that $body(r)^+ = \{a_1, \ldots, a_m\}$ and $body(r)^- = \{a_{m+1}, \ldots, a_n\}$. A rule r is said to be positive, if $body(r)^- = \emptyset$. Accordingly, a program is called positive, if all its rules are positive. The set of atoms occurring in a logic program P is denoted by $atom(P)$, and $body(P) = \{body(r) \mid r \in P\}$ is the set of all bodies of rules in P. For rule bodies sharing the same head a, we define furthermore $body_P(a) = \{body(r) \mid r \in P, head(r) = a\}$.

A set $X \subseteq \mathcal{A}$ of ground atoms is a *model* of a propositional logic program P, if $head(r) \in X$ whenever $body(r)^+ \subseteq X$ and $body(r)^- \cap X = \emptyset$ for every $r \in P$. For instance, Program P_1 has six models, among which we find $\{a, c\}$ and $\{a, b, c, d\}$.

In ASP, the semantics of P is given by its *stable models* (Gelfond and Lifschitz, 1988). To this end, the *reduct*, P^X, of P relative to a set X of atoms is defined by

$$P^X = \{head(r) \leftarrow body(r)^+ \mid r \in P, body(r)^- \cap X = \emptyset\}.$$

---

[1]As common in Database Systems and Logic Programming, default negation refers to the absence of information, while "classical" negation stipulates the presence of the negated information, or informally, ${\sim}a$ stands for $a \notin X$ while $\neg a$ requires $\neg a \in X$ for some interpretation X (see also Section 2.3.4).

Note that P^X is a positive program, thus possessing a unique $\subseteq$-minimal model. Given this, X is a *stable model* of P, if X is the $\subseteq$-minimal model of P^X.

For illustration, consider the two aforementioned models of P_1, namely $\{a, c\}$ and $\{a, b, c, d\}$, and check whether they are stable.

X	P_1^X			$\subseteq$ -minimal model of P_1^X
$\{a, c\}$	$P_1^{\{a,c\}}$	$=$	$\{a \leftarrow,\ c \leftarrow\}$	$\{a, c\}$
$\{a, b, c, d\}$	$P_1^{\{a,b,c,d\}}$	$=$	$\{a \leftarrow\}$	$\{a\}$

We observe that $\{a, c\}$ is indeed a stable model of P_1, while $\{a, b, c, d\}$ is not.

Pragmatically, the reduct P^X of a program P relative to a set X of atoms is obtained by

1. deleting each rule having a negative literal $\sim a$ in its body with $a \in X$ and then

2. eliminating all negative literals of the form $\sim a$ in the bodies of the remaining rules.

This view emphasizes that atoms preceded by '$\sim$' are evaluated in the traditional way with respect to the model candidate X.

We can rephrase the definition of a stable model in a more compact way by using the (consequence) operator Cn to yield the smallest model of a positive program. With it, the stable models, X, of a logic program P can be characterized as fixpoints of the equation $Cn(P^X) = X$. This characterization intuitively reflects that X is *stable* under "applying rules from P." While all atoms in the model $\{a, c\}$ of P_1 are stable under applying the rules in P_1, in $\{a, b, c, d\}$ the truth of the atoms b, c, and d cannot be justified. In this sense, each atom in a stable model is "provable" by rules from P, hinting at the above mentioned constructive flavor of ASP. Hence, negative literals must only be true, while positive ones must also be provable. This informal observation is made precise in Chapter 5; see also Section 2.4 below.

Note that any stable model of P is also a ($\subseteq$-minimal) model of P, whereas the converse does not hold in general. A positive program has a unique stable model, given by its smallest model. Also, for all stable models X and Y of a normal program, we have $X \not\subset Y$. More pragmatically, we note that $X \subseteq Cn(P^X) \subseteq head(P^X)$, that is, stable models are formed from programs' heads. In view of this, $\{a, b, c, d\}$ cannot be a stable model of P_1 because it already fails to be a subset of $head(P_1)$.

Let us use the previous characterization to show that a logic program may have zero, one, or multiple stable models. To see this, consider the following three examples inspecting the respective set of candidate models given as traditional truth tables.

The first program, P_2, has the single stable model $\{b\}$.

$$P_2 = \{a \leftarrow a,\ b \leftarrow \sim a\}$$

This program has three models, $\{a\}$, $\{b\}$, and $\{a, b\}$. The first and last model are not stable because neither admits a "non-circular derivation" of a from the reduct $P_2^{\{a\}}$ and $P_2^{\{a,b\}}$, respectively. To see this, let us consider in turn all candidate sets, $X \subseteq atom(P_2)$, the resulting reducts, P_2^X, and their smallest models, $Cn(P_2^X)$.

X	P_2^X				$Cn(P_2^X)$		
$\emptyset$	$P_2^{\emptyset}$	$=$	$\{a \leftarrow a, \ b \leftarrow$	$\}$	$Cn(P_2^{\emptyset})$	$=$	$\{b\}$
$\{a\}$	$P_2^{\{a\}}$	$=$	$\{a \leftarrow a$	$\}$	$Cn(P_2^{\{a\}})$	$=$	$\emptyset$
$\{b\}$	$P_2^{\{b\}}$	$=$	$\{a \leftarrow a, \ b \leftarrow$	$\}$	$Cn(P_2^{\{b\}})$	$=$	$\{b\}$
$\{a, b\}$	$P_2^{\{a,b\}}$	$=$	$\{a \leftarrow a$	$\}$	$Cn(P_2^{\{a,b\}})$	$=$	$\emptyset$

We see that among all four candidate sets only $\{b\}$ satisfies the equation $\{b\} = Cn(P_2^{\{b\}})$, and it thus constitutes the only stable model of P_2.

Next, consider Program P_3, whose two rules can be interpreted as describing a choice between a and b.

$$P_3 \ = \ \{a \leftarrow \sim b, \ b \leftarrow \sim a\}$$

As with P_2, Program P_3 has three models, $\{a\}$, $\{b\}$, and $\{a, b\}$. Unlike the above, the first two among them are stable.

X	P_3^X					$Cn(P_3^X)$		
$\emptyset$	$P_3^{\emptyset}$	$=$	$\{a \leftarrow$	$, \ b \leftarrow$	$\}$	$Cn(P_3^{\emptyset})$	$=$	$\{a, b\}$
$\{a\}$	$P_3^{\{a\}}$	$=$	$\{a \leftarrow$		$\}$	$Cn(P_3^{\{a\}})$	$=$	$\{a\}$
$\{b\}$	$P_3^{\{b\}}$	$=$	$\{$	$b \leftarrow$	$\}$	$Cn(P_3^{\{b\}})$	$=$	$\{b\}$
$\{a, b\}$	$P_3^{\{a,b\}}$	$=$	$\{$		$\}$	$Cn(P_3^{\{a,b\}})$	$=$	$\emptyset$

Pairs of rules as in P_3 provide us with non-deterministic language constructs and form the basis of choice rules, introduced in Section 2.3.2.

The two previous examples suggest that stable models take the format of rules into account. Although both P_2 and P_3 have the same models, the different format of the program induces different "derivations" which in turn lead to different stable models.

Finally, we give a program admitting no stable model.

$$P_4 \ = \ \{a \leftarrow \sim a\}$$

Although, P_4 lacks a stable model, it has a single model containing a.

X	P_4^X				$Cn(P_4^X)$		
$\emptyset$	$P_3^{\emptyset}$	$=$	$\{a \leftarrow$	$\}$	$Cn(P_3^{\emptyset})$	$=$	$\{a\}$
$\{a\}$	$P_3^{\{a\}}$	$=$	$\{$	$\}$	$Cn(P_3^{\{a\}})$	$=$	$\emptyset$

Rules as in P_4 play an important practical role in implementing integrity constraints, as detailed in Section 2.3.2.

2.3   LANGUAGE EXTENSIONS

After the introduction to the basic propositional fragment of ASP, we now successively extend the language of ASP while focusing on the *smodels* format. We start with a brief account of first-order variables in ASP and show how they can be (semantically) eliminated. In the remainder, we then concentrate on the propositional case. For this, we presuppose from Section 2.3.2 on that $\mathcal{A}$ is a set of ground atoms and often leave it implicit. The remaining sections 2.3.2 to 2.3.5 introduce the salient language constructs constituting ASP's modeling language.

2.3.1   FIRST-ORDER VARIABLES

Following the tradition of Logic Programming, we view rules with first-order variables as schemes representing their sets of ground instances. To be more precise, a rule is ground, if it contains no variables. The set $grd(r)$ of *ground instances* of a rule r is the set of all ground rules obtained by replacing all variables in r by ground terms. Accordingly, the ground instantiation of a program P is given by $grd(P) = \bigcup_{r \in P} grd(r)$.

For instance, the program, P,

```
arc(1,1).
arc(1,2).
edge(X,Y) :- arc(X,Y), arc(Y,X).
```

yields the ground program, $grd(P)$,

```
arc(1,1).
arc(1,2).
edge(1,1) :- arc(1,1), arc(1,1).
edge(2,2) :- arc(2,2), arc(2,2).
edge(1,2) :- arc(1,2), arc(2,1).
edge(2,1) :- arc(2,1), arc(1,2).
```

Note that the last three ground rules are superfluous and can thus be removed from $grd(P)$. That is, the above ground program has the same stable model as the first three ground rules, comprising $arc(1,1), arc(1,2)$, and $edge(1,1)$. The effective computation of compact representations of $grd(P)$ is the subject of Chapter 4.

The semantics of programs with variables is then a direct extension of the propositional case. Given a (normal) logic program P over a set of (non-ground) atoms $\mathcal{A}$, a set $X \subseteq grd(\mathcal{A})$ of ground atoms is a stable model of P, if X is the $\subseteq$-minimal model of $grd(P)^X$.

Given that a program P with variables can be regarded as an abbreviation for $grd(P)$, we henceforth concentrate on the propositional case.

2.3.2   CORE LANGUAGE

This section introduces the essential modeling language of ASP, sufficient for expressing all search problems in *NP*. To this end, we mainly focus on logic programs consisting of

- normal rules,

- choice rules,

- cardinality rules, and

- weight rules.

Together with optimization statements introduced in the following section, this collection constitutes the basic language constructs accepted by ASP solvers like *smodels* and *clasp* (see also Section 7.1.4). Except for optimization statements, all language extensions are complexity-preserving. Hence, we generally fix their meaning via translations reducing them to normal logic programs rather than providing genuine semantics.

A compact formal account on the core language of ASP (extended by disjunction) is given in Appendix A.

Integrity constraints   An integrity constraint is of the form

$$\leftarrow a_1, \ldots, a_m, \sim a_{m+1}, \ldots, \sim a_n \tag{2.2}$$

where $0 \leq m \leq n$ and each a_i is an atom for $1 \leq i \leq n$.

An integrity constraint rules out stable models satisfying its body literals. Their purpose is to eliminate unwanted solution candidates. No atoms are derivable through integrity constraints. For instance, the integrity constraint

```
:- edge(3,7), color(3,red), color(7,red).
```

can be used in a graph coloring problem to express that vertices 3 and 7 must not both be colored red if they are connected.

An integrity constraint can be translated into a normal rule. To this end, the constraint in (2.2) is mapped onto the rule

$$x \leftarrow a_1, \ldots, a_m, \sim a_{m+1}, \ldots, \sim a_n, \sim x$$

where x is a new symbol, that is, $x \notin \mathcal{A}$.

To illustrate this, let us extend P_3 from Page 15 by integrity constraints as follows.

$$
\begin{aligned}
P_3 \cup \{\leftarrow a\} &= \{a \leftarrow \sim b, \ b \leftarrow \sim a\} \cup \{\leftarrow a\} \\
P_3 \cup \{\leftarrow \sim a\} &= \{a \leftarrow \sim b, \ b \leftarrow \sim a\} \cup \{\leftarrow \sim a\}
\end{aligned}
$$

From the two stable models of P_3, the first program only admits $\{b\}$, while the second one yields $\{a\}$. The same stable models are obtained from Program $P_3 \cup \{x \leftarrow a, \sim x\}$ and $P_3 \cup \{x \leftarrow \sim a, \sim x\}$, respectively.

In general, the addition of integrity constraints to a logic program can neither produce new stable models nor alter existing ones; rather it can only lead to their elimination.

Choice rules   A choice rule is of the form[2]

$$\{a_1, \ldots, a_m\} \leftarrow a_{m+1}, \ldots, a_n, \sim a_{n+1}, \ldots, \sim a_o \qquad (2.3)$$

where $0 \le m \le n \le o$ and each a_i is an atom for $1 \le i \le o$.

The idea of a choice rule is to express choices over subsets of atoms. Any subset of its head atoms can be included in a stable model, provided the body literals are satisfied. Thus, for instance, the program $P = \{a \leftarrow, \{b\} \leftarrow a\}$ has two stable models, $\{a\}$ and $\{a, b\}$. For another example, at a grocery store you may or may not buy pizza, wine, or corn.

```
{ buy(pizza), buy(wine), buy(corn) } :- at(grocery).
```

A choice rule of form (2.3) can be translated into $2m + 1$ rules

$$a' \leftarrow a_{m+1}, \ldots, a_n, \sim a_{n+1}, \ldots, \sim a_o$$

$$a_1 \leftarrow a', \sim \overline{a_1} \quad \ldots \quad a_m \leftarrow a', \sim \overline{a_m}$$
$$\overline{a_1} \leftarrow \sim a_1 \quad \ldots \quad \overline{a_m} \leftarrow \sim a_m$$

by introducing new atoms $a', \overline{a_1}, \ldots, \overline{a_m}$. Applying this transformation to the choice rule $\{b\} \leftarrow a$ in our example program P yields

$$a \leftarrow \qquad b' \leftarrow a$$
$$b \leftarrow b', \sim \overline{b}$$
$$\overline{b} \leftarrow \sim b$$

This program has two stable models, $\{a, b', \overline{b}\}$ and $\{a, b', b\}$, whose intersections with the atoms in the original program correspond to the stable models indicated above.

Cardinality rules   A cardinality rule is of the form

$$a_0 \leftarrow l \{ a_1, \ldots, a_m, \sim a_{m+1}, \ldots, \sim a_n \} \qquad (2.4)$$

where $0 \le m \le n$ and each a_i is an atom for $0 \le i \le n$; l is a non-negative integer.

Cardinality rules allow for controlling the cardinality of subsets of atoms via the lower bound l. That is, the head atom belongs to a stable model, if the latter satisfies at least l body literals. For example, Program $P = \{a \leftarrow, c \leftarrow 1 \{a, b\}\}$ has the stable model $\{a, c\}$. Here is a less artificial example of a cardinality rule, describing that one passes Course 42, provided one passes two out of three assignments.

[2]The inclusion of default negated literals among the head literals of plain choice rules is without effect. No matter whether a literal $\sim a$ is chosen or not, it cannot contribute to a stable model.

```
pass(c42) :- 2 { pass(a1), pass(a2), pass(a3) }.
```

Also, cardinality rules can be translated into normal programs. To this end, we replace a rule of form (2.4) by

$$a_0 \leftarrow ctr(1, l)$$

where atom $ctr(i, j)$ represents the fact that at least j of the literals having an equal or greater index than i, are in a stable model.

The definition of $ctr/2$ is given for $0 \leq k \leq l$ by the rules

$$
\begin{aligned}
ctr(i, k{+}1) &\leftarrow ctr(i + 1, k), a_i \\
ctr(i, k) &\leftarrow ctr(i + 1, k) \qquad \text{for } 1 \leq i \leq m \\[6pt]
ctr(j, k{+}1) &\leftarrow ctr(j + 1, k), {\sim} a_j \\
ctr(j, k) &\leftarrow ctr(j + 1, k) \qquad \text{for } m + 1 \leq j \leq n \\[6pt]
ctr(n + 1, 0) &\leftarrow
\end{aligned}
$$

For illustration, let us apply this transformation to the cardinality rule $c \leftarrow 1 \{a, b\}$ in the last program P. We get the following program.

$$
\begin{aligned}
a &\leftarrow & c &\leftarrow ctr(1, 1) \\
& & ctr(1, 2) &\leftarrow ctr(2, 1), a \\
& & ctr(1, 1) &\leftarrow ctr(2, 1) \\
& & ctr(2, 2) &\leftarrow ctr(3, 1), b \\
& & ctr(2, 1) &\leftarrow ctr(3, 1) \\
& & ctr(1, 1) &\leftarrow ctr(2, 0), a \\
& & ctr(1, 0) &\leftarrow ctr(2, 0) \\
& & ctr(2, 1) &\leftarrow ctr(3, 0), b \\
& & ctr(2, 0) &\leftarrow ctr(3, 0) \\
& & ctr(3, 0) &\leftarrow
\end{aligned}
$$

This program yields the stable model $\{a, ctr(3, 0), ctr(2, 0), ctr(1, 0), ctr(1, 1), c\}$. Reduced to the original atoms, we thus get $\{a, c\}$.

Note that unlike the above translations, the one for cardinality rules is quadratic in space. This is why many ASP solvers employ a dedicated treatment once the number of literals gets too large (see Section 7.2).

Interestingly, cardinality rules could be used as an alternative base construct instead of normal rules. To see this, observe that any normal rule of form (2.1) can be expressed as a cardinality rule of the form

$$a_0 \leftarrow n \{ a_1, \ldots, a_m, {\sim} a_{m+1}, \ldots, {\sim} a_n \}.$$

Next, we generalize cardinality rules for expressing more general forms of cardinality constraints, and we provide semantics via translations into simpler forms of programs.

At first, we consider cardinality rules with upper bounds.

$$a_0 \leftarrow l \{ a_1, \ldots, a_m, \sim a_{m+1}, \ldots, \sim a_n \} u \tag{2.5}$$

Such rules extend the syntax of cardinality rules in (2.4) by adding another non-negative integer, u, serving as an upper bound on the cardinality of the satisfied body literals. The single constraint in the body of (2.5) is commonly referred to as a *cardinality constraint*.

A cardinality rule with an upper bound can be expressed by the following three rules (introducing new symbols b and c).

$$
\begin{aligned}
a_0 &\leftarrow b, \sim c \\
b &\leftarrow l \{ a_1, \ldots, a_m, \sim a_{m+1}, \ldots, \sim a_n \} \\
c &\leftarrow u+1 \{ a_1, \ldots, a_m, \sim a_{m+1}, \ldots, \sim a_n \}
\end{aligned}
\tag{2.6}
$$

So far, all cardinality constraints occurred in rule bodies only. We next consider rules with cardinality constraints as heads. Such a rule is of the form

$$l \{a_1, \ldots, a_m, \sim a_{m+1}, \ldots, \sim a_n\} u \leftarrow a_{n+1}, \ldots, a_o, \sim a_{o+1}, \ldots, \sim a_p \tag{2.7}$$

where $0 \leq m \leq n \leq o \leq p$ and each a_i is an atom for $1 \leq i \leq p$; l and u are non-negative integers. For example, we can express that vertex 42 must be colored with exactly one color, among red, green, and blue, as follows.

```
1 { color(v42,red), color(v42,green), color(v42,blue) } 1 :- vertex(v42).
```

Rule (2.7) amounts to an (extended) choice rule whose selection is limited by the associated lower and upper bound. This is also reflected by its transform. Making this precise, a rule of form (2.7) can be represented as follows.

$$
\begin{aligned}
b &\leftarrow a_{n+1}, \ldots, a_o, \sim a_{o+1}, \ldots, \sim a_p \\
\{a_1, \ldots, a_m\} &\leftarrow b \\
c &\leftarrow l \{a_1, \ldots, a_m, , \sim a_{m+1}, \ldots, \sim a_n\} u \\
&\leftarrow b, \sim c
\end{aligned}
\tag{2.8}
$$

The first two rules give the choice rule obtained by dropping both bounds in (2.7); also negative head literals are dropped because they do not give rise to deriving any atoms. In contrast, atoms occurring positively in the head can be derived by choice. The third rule checks whether the selection of head atoms respects the cardinality imposed by l and u. Finally, the integrity constraint eliminates selections invalidating the bounds.

At last, let us combine the above translations and consider full-fledged cardinality rules, in the sense that cardinality constraints can be used instead of atoms in normal rules. This leads us to rules of the form

$$l_0 \ S_0 \ u_0 \leftarrow l_1 \ S_1 \ u_1, \ldots, l_n \ S_n \ u_n \tag{2.9}$$

where for $0 \leq i \leq n$ each $l_i \, S_i \, u_i$ is a cardinality constraint. Note that a normal rule as in (2.1) can be expressed in the form of (2.9) as

$$1 \{a_0\} \leftarrow 1 \{a_1\}, \ldots 1 \{a_m\}, \{a_{m+1}\} \, 0, \ldots, \{a_n\} \, 0.$$

Alternatively, we may express $\{a_i\} \, 0$ as $1 \{\sim a_i\}$ for $m < i \leq n$.

Such a general rule as in (2.9) can be represented by the following rules for $0 \leq i \leq n$.

$$
\begin{aligned}
a &\leftarrow b_1, \ldots, b_n, \sim c_1, \ldots, \sim c_n & &\leftarrow a, \sim b_0 & b_i &\leftarrow l_i \, S_i \\
& & &\leftarrow a, c_0 & c_i &\leftarrow u_i{+}1 \, S_i \\
& & S_0{}^+ &\leftarrow a
\end{aligned}
$$

where a, b_i, c_i are fresh symbols not appearing in the underlying set of atoms $\mathcal{A}$.

Weight rules A weight rule is defined in analogy to (2.4) as a rule of form

$$a_0 \leftarrow l \{ a_1 = w_1, \ldots, a_m = w_m, \sim a_{m+1} = w_{m+1}, \ldots, \sim a_n = w_n \} \qquad (2.10)$$

where $0 \leq m \leq n$ and each a_i is an atom for $0 \leq i \leq n$; and l and w_i are integers for $1 \leq i \leq n$. A weighted literal, $\ell_i = w_i$, associates each literal ℓ_i with a weight w_i.

The meaning of weight rules along with the resulting notions of weight constraints can be given in analogy to cardinality rules and the various forms of rules including cardinality constraints, respectively. The major difference lies in extending the definition of the *ctr/2* predicate. Rather than incrementing counters by one, the weight of the respective literal must be added. All remaining definitions follow analogously.

A more direct perspective on such language extensions is given by their satisfaction. For example, a cardinality constraint $l \{ a_1, \ldots, a_m, \sim a_{m+1}, \ldots, \sim a_n \} u$ is satisfied by an interpretation X, if the number of its literals belonging to X is between l and u (inclusive), in other words, if $l \leq |\, (\{a_1, \ldots, a_m\} \cap X) \cup (\{a_{m+1}, \ldots, a_n\} \setminus X) \,| \leq u$. Obviously, cardinality constraints can be regarded as special weight constraints, in which all literals have weight 1.

Accordingly, a general *weight constraint* is of the form

$$l \{ a_1 = w_1, \ldots, a_m = w_m, \sim a_{m+1} = w_{m+1}, \ldots, \sim a_n = w_n \} u \qquad (2.11)$$

where $0 \leq m \leq n$ and each a_i is an atom for $1 \leq i \leq n$; and l, u and w_i are integers for $1 \leq i \leq n$. The meaning of a weight constraint can also be captured through its satisfaction by a stable model: a weight constraint as in (2.11) is satisfied by an interpretation X, if

$$l \leq \left(\sum_{1 \leq i \leq m, a_i \in X} w_i + \sum_{m < i \leq n, a_i \notin X} w_i \right) \leq u.$$

This definition stresses the fact that cardinality and weight constraints amount to constraints on count and sum aggregate functions. Note that in general the meaning of a weight constraint is defined via an appropriate reduct. A crisp formal account on weight constraints is given in Appendix A.

Here is an example of a weight constraint about choosing courses providing a certain number of credits.

```
10 { course(db)=6, course(ai)=6, course(project)=8, course(xml)=3 } 20
```

Although weight constraints may look like a simple generalization of cardinality constraints, the possibility of including both positive and negative weights for atoms can lift computational complexity by one level in the polynomial time hierarchy. Such elevated complexity is due to the loss of "constructiveness" in the verification of stable models, where the derivation of an atom with negative (or positive) weight may undercut a formerly established lower (or upper) bound.

See also Section 7.1 on the treatment of aggregates in the ASP grounder *gringo*.

Conditional literals A conditional literal is of the form $\ell : \ell_1 : \cdots : \ell_n$ for $0 \leq i \leq n$. The purpose of this simple yet powerful language construct is to govern the instantiation of the "head literal" ℓ through the literals $\ell_1, \ldots, \ell_n$. In this respect, a conditional literal $\ell : \ell_1 : \cdots : \ell_n$ can be regarded as the list of elements in the set $\{\ell \mid \ell_1, \ldots, \ell_n\}$.

For example, given three facts color(red), color(green), and color(blue), the conditional literal in the cardinality constraint

```
1 { color(v42,C) : color(C) } 1 :- vertex(v42).
```

expands to the cardinality constraint

```
1 { color(v42,red), color(v42,green), color(v42,blue) } 1 :- vertex(v42).
```

However, the final form of the expanded conditional literal is context-dependent. For instance, given the above facts, the integrity constraint

```
:- color(v42,C) : color(C).
```

results in

```
:- color(v42,red), color(v42,green), color(v42,blue).
```

Similarly, conditional literals can be used in optimization statements (see Section 2.3.3) and disjunctive rule heads (see Section 2.3.5). A sophisticated use of conditional literals in various contexts is shown in Listing 4.7 on Page 64 as well as Listing 8.10 on Page 163.

2.3.3 OPTIMIZATION STATEMENTS

For solving (multi-criteria) optimization problems, ASP allows for expressing cost functions subject to minimization and/or maximization. Such objective functions are expressed in ASP in terms of optimization statements. In fact, maximization is defined analogously to minimization, so that we concentrate on the latter in the sequel.

A *minimize statement* is of the following form:

$$minimize\{ \ell_1 = w_1 @ p_1, \ldots, \ell_n = w_n @ p_n \}. \tag{2.12}$$

As with weight constraints, every ℓ_i is a literal (that is, of form a_i or $\sim a_i$) and every w_i an integer weight for $1 \leq i \leq n$; in addition, p_i provides an integer priority level. Priorities allow for representing lexicographically ordered minimization objectives, greater levels being more significant than smaller ones. A maximize statement of the form $maximize\{\ell_1 = w_1 @ p_1, \ldots, \ell_n = w_n @ p_n\}$ can be represented by the minimize statement $minimize\{\ell_1 = -w_1 @ p_1, \ldots, \ell_n = -w_n @ p_n\}$.

A minimize statement is a directive that instructs the ASP solver to compute optimal stable models by minimizing a weighted sum of elements. For example, when configuring a computer, we may want to maximize hard disk capacity, while minimizing price.[3]

```
#maximize[ hd(1)=250@1, hd(2)=500@1, hd(3)=750@1, hd(4)=1000@1 ].
#minimize[ hd(1)=30@2,  hd(2)=40@2,  hd(3)=60@2,  hd(4)=80@2   ].
```

The priority levels indicate that (minimizing) price is more important than (maximizing) capacity. We observe that in practice minimize (and maximize) statements are preceded by # in order to indicate that they are directives and do not belong to the program as such.

A minimize statement distinguishes optimal stable models of a program P in the following way. For any $X \subseteq \mathcal{A}$ and integer p, let Σ_p^X denote the sum of weights w over all occurrences of weighted literals $\ell = w @ p$ in (2.12) such that ℓ is satisfied by X. A stable model X of P is dominated if there is a stable model Y of P such that $\Sigma_p^Y < \Sigma_p^X$ and $\Sigma_{p'}^Y = \Sigma_{p'}^X$ for all $p' > p$, and optimal otherwise.

2.3.4   TWO (AND A HALF) KINDS OF NEGATION

The addition of a second kind of negation, resembling classical negation, is mainly motivated by a desire to ease knowledge representation. Pragmatically, the introduction of negation amounts to the addition of new language symbols for all atoms, along with the addition of rules fixing the relation of these new atoms to their original counterparts.

Given that an atom a is satisfied by a stable model whenever $a \in X$, the difference between a classically negated literal $\neg a$ and a default negated one, $\sim a$, intuitively boils down to the difference between $\neg a \in X$ and $a \notin X$, respectively. A popular example illustrating this distinction is given by the two programs

$$P = \{cross \leftarrow \sim train\} \quad \text{and} \quad P' = \{cross \leftarrow \neg train\}.$$

While informally the first program suggests crossing the tracks whenever we *do not know* whether a train approaches, the second one advises doing so whenever we *know* that no train arrives. Accordingly, P has the stable model $\{cross\}$, whereas P' yields the empty stable model. We must add the fact $\neg train \leftarrow$ to P' to obtain the same conclusion.

To make things precise, we extend our set of atoms $\mathcal{A}$ by $\overline{\mathcal{A}} = \{\neg a \mid a \in \mathcal{A}\}$ such that $\mathcal{A} \cap \overline{\mathcal{A}} = \emptyset$. That is, $\neg a$ is the classical negation of a and vice versa. The semantics of classical negation is

---

[3]In *gringo*, optimization statements deal with multisets, enclosed in brackets like [and].

enforced by the addition of the following set of rules[4]

$$P^\neg = \{a \leftarrow b, \neg b \mid a \in (\mathcal{A} \cup \overline{\mathcal{A}}), b \in \mathcal{A}\}.$$

For illustration, let us extend once more Program P_3 from Page 15.

$$P_3 \cup \{c \leftarrow b, \ \neg c \leftarrow b\} = \{a \leftarrow \sim b, \ b \leftarrow \sim a\} \cup \{c \leftarrow b, \ \neg c \leftarrow b\}.$$

The resulting program, viz. $P_3 \cup \{c \leftarrow b, \ \neg c \leftarrow b\} \cup P^\neg$, has a single stable model $\{a\}$. Note that the second stable model $\{b\}$ of P_3 is eliminated by the contradiction, c and $\neg c$, obtained from b.

Strictly speaking, the stable models obtained from programs with classical negation are no models because a stable model of the transformed program may contain an atom and its negation. For instance, the program $P_3 \cup \{c \leftarrow, \ \neg c \leftarrow\} \cup P^\neg$ yields the stable model $\mathcal{A} \cup \overline{\mathcal{A}}$. In fact, a transformed program has either only stable models free of complementary literals or the single stable model $\mathcal{A} \cup \overline{\mathcal{A}}$.

Finally, the appearance of default negation in rule heads can be reduced to normal programs. To this end, we must also extend our set $\mathcal{A}$ of atoms by $\widetilde{\mathcal{A}} = \{\tilde{a} \mid a \in \mathcal{A}\}$ such that $\mathcal{A} \cap \widetilde{\mathcal{A}} = \emptyset$. The stable models of a program P with default negation in the head are then provided by the stable models (projected to $\mathcal{A}$) of the following normal program.

$$\begin{aligned}
\widetilde{P} = \ & \{r \in P \mid head(r) \neq \sim a\} \\
& \cup \{\leftarrow body(r) \cup \{\sim \tilde{a}\} \mid r \in P, head(r) = \sim a\} \\
& \cup \{\tilde{a} \leftarrow \sim a \mid r \in P, head(r) = \sim a\}.
\end{aligned} \qquad (2.13)$$

This translation also works for disjunctive logic programs, as illustrated below.

2.3.5   TRUE LANGUAGE EXTENSIONS

Finally, let us deal with language constructs that cannot be reduced to normal logic programs because they allow for capturing problems in NP^{NP} (see Section 2.4.3 below). For brevity, however, we restrict ourselves to brief introductions to the syntax and semantics of disjunctive logic programs and general propositional theories.

A disjunctive logic program consists of rules of the form

$$a_1 \ ; \ldots ; \ a_m \leftarrow a_{m+1}, \ldots, a_n, \sim a_{n+1}, \ldots, \sim a_o \qquad (2.14)$$

where $0 \leq m \leq n \leq o$ and each a_i is an atom for $1 \leq i \leq o$.

For example, we can express that vertex 42 must be colored with one color, among red, green, and blue, as follows.

```
color(v42,red) | color(v42,green) | color(v42,blue)   :- vertex(v42).
```

[4]Existing ASP systems implement a different semantics by setting $P^\neg$ to $\{ \leftarrow b, \neg b \mid b \in \mathcal{A}\}$. This eliminates the putative "stable model" $\mathcal{A} \cup \overline{\mathcal{A}}$.

Alternatively, this rule can be written by using conditional literals in the following way.

```
color(v42,C) : color(C) :- vertex(v42).
```

For a disjunctive rule r as in (2.14), let us redefine $head(r) = \{a_1, \ldots, a_m\}$ (cf. Section 2.2). A set $X \subseteq \mathcal{A}$ of ground atoms is a *model* of a propositional logic program P, if $head(r) \cap X \neq \emptyset$ whenever $body(r)^+ \subseteq X$ and $body(r)^- \cap X = \emptyset$ for every $r \in P$. Given this, a set X of ground atoms is a *stable model* of P, if X is some $\subseteq$-minimal model of P^X.

Note that the reduct of a disjunctive program may have several minimal models. Furthermore observe that disjunction in ASP is neither strictly inclusive nor exclusive but subject to minimization. To see this, consider the following positive program.

$$P_5 = \{a ; b \leftarrow\}$$

Among the three models of the program, only $\{a\}$ and $\{b\}$ are $\subseteq$-minimal and thus stable models of P_5. Adding the facts $\{a \leftarrow, b \leftarrow\}$ to P_5 yields a single stable model $\{a, b\}$. An inclusive disjunction can be modeled with a cardinality rule as in (2.7) by setting $l = 1$ and $u = n$; similarly, an exclusive disjunction is obtained by setting $l = 1$ and $u = 1$ in (2.7).

In view of the translation in (2.13), we may even consider disjunctive rules of the form

$$a_1 ; \ldots ; a_m ; \sim a_{m+1} ; \ldots ; \sim a_n \leftarrow a_{n+1}, \ldots, a_o, \sim a_{o+1}, \ldots, \sim a_p$$

where $0 \leq m \leq n \leq o \leq p$ and each a_i is an atom for $1 \leq i \leq p$. Such rules can be reduced to disjunctive rules of form (2.14) by adapting the translation in (2.13).[5]

$$\begin{aligned} \widetilde{P} = \quad &\{head(r)^+ \leftarrow body(r) \cup \{\sim\tilde{a} \mid a \in head(r)^-\} \mid r \in P\} \\ &\cup \{\tilde{a} \leftarrow \sim a \mid r \in P, a \in head(r)^-\}. \end{aligned}$$

For illustration, consider the following program.

$$P_6 = \{a ; \sim a \leftarrow\}$$

The extended translation from (2.13) results in the disjunctive program

$$\widetilde{P}_6 = \{a \leftarrow \sim\tilde{a}\} \cup \{\tilde{a} \leftarrow \sim a\}.$$

Program $\widetilde{P}_6$ has two stable models, $\{a\}$ and $\{\tilde{a}\}$, whose projection on $\mathcal{A}$, namely $\{a\}$ and $\emptyset$, gives the stable models of P_6.

Finally, let us look at propositional theories under stable models semantics. The satisfaction relation $X \models \phi$ between a set X of atoms and a (set of) formula(s) ϕ is defined as in propositional logic (treating $\sim$ like $\neg$). The reduct, ϕ^X, of a formula ϕ relative to a set X of atoms is defined

---

[5]We slightly abuse notation and represent a disjunctive head in set notation.

recursively as follows.

$$\phi^X = \bot \qquad \text{if } X \not\models \phi$$
$$\phi^X = \phi \qquad \text{if } \phi \in X$$
$$\phi^X = (\psi^X \circ \mu^X) \quad \text{if } X \models \phi \text{ and } \phi = (\psi \circ \mu) \text{ for } \circ \in \{\wedge, \vee, \rightarrow\}$$
$$\phi^X = \top \qquad \text{if } X \not\models \psi \text{ and } \phi = \sim\psi$$

The reduct, Φ^X, of a propositional theory Φ relative to a set X of atoms is defined as $\Phi^X = \{\phi^X \mid \phi \in \Phi\}$. Note that the reduct is a positive formula, that is, it is free of negation. As above, a set X of ground atoms is a *stable model* of Φ, if X is some $\subseteq$-minimal model of Φ^X.

Any stable model of Φ is a also a model of Φ. However, stable models are in general not minimal models. Interestingly, if X is a stable model of Φ, then it is the smallest model of Φ^X. This is due to the stronger reduct, whose evaluation is not limited to negative formulas. As an example, consider $a \vee b$. We get $(a \vee b)^{\{a\}} = a \vee \bot$ whose only minimal model is $\{a\}$, establishing the fact that $\{a\}$ is a stable model of $a \vee b$. Similarly, we get that $\{b\}$ is the second stable model of $a \vee b$ (cf. Program P_5 above). For another example, consider $a \vee \sim a$. We get $(a \vee \sim a)^{\{a\}} = a \vee \bot$ and thus $\{a\}$ as stable model. Analogously, we obtain $(a \vee \sim a)^{\emptyset} = \bot \vee \top$ yielding the empty stable model (cf. Program P_6 above). Unlike this, the formula $\sim\sim a \vee \sim a$ admits the empty stable model $\emptyset$ only. To see this, observe that $\emptyset$ is the smallest model of both $(\sim\sim a \vee \sim a)^{\{a\}} = \top \vee \bot$ and $(\sim\sim a \vee \sim a)^{\emptyset} = \bot \vee \top$, respectively. This nicely illustrates that double negated literals are not equivalent to their base atoms.

Finally, we note that under the stable models semantics any propositional theory can be reduced to an equivalent set of disjunctive rules of form (2.14).

2.4   COMPUTATIONAL ASPECTS

To begin with, we develop an initial approach to ASP solving from the definition of stable models in the context of normal logic programs. The next two sections provide brief summaries of ASP's reasoning modes and basic complexity results.

2.4.1   COMPUTATION FROM FIRST PRINCIPLES

We begin with a brief development of an algorithm for computing stable models in terms of the preceding formal concepts.

The smallest model, $Cn(P)$, of a positive program P can be computed via its associated *consequence operator* T_P. For a set of atoms X, we define

$$T_P X = \{head(r) \mid r \in P \text{ and } body(r) \subseteq X\}. \tag{2.15}$$

Iterated applications of T_P are written as T_P^j for $j \geq 0$, where

$$T_P^0 X = X \quad \text{and} \quad T_P^i X = T_P T_P^{i-1} X \text{ for } i \geq 1.$$

For any positive program P, we have $Cn(P) = \bigcup_{i \geq 0} T_P^i \emptyset$. Since T_P is monotonic, $Cn(P)$ is the smallest fixpoint of T_P.

For illustration, consider the following positive program

$$P = \{\, a \leftarrow, \quad b \leftarrow a, \quad c \leftarrow a, b, \quad e \leftarrow f \,\}$$

along with the computation of its consequences, $Cn(P)$, via iterated applications of T_P.

$$
\begin{aligned}
T_P^0 \emptyset &= \emptyset \\
T_P^1 \emptyset &= \{a\} &= T_P T_P^0 \emptyset &= T_P \emptyset \\
T_P^2 \emptyset &= \{a, b\} &= T_P T_P^1 \emptyset &= T_P \{a\} \\
T_P^3 \emptyset &= \{a, b, c\} &= T_P T_P^2 \emptyset &= T_P \{a, b\} \\
T_P^4 \emptyset &= \{a, b, c\} &= T_P T_P^3 \emptyset &= T_P \{a, b, c\}
\end{aligned}
$$

To see that $Cn(P) = \{a, b, c\}$ is the smallest fixpoint of T_P, note that $T_P\{a, b, c\} = \{a, b, c\}$ and $T_P X \neq X$ for every $X \subset \{a, b, c\}$.

The computation of the smallest model $Cn(P^X)$ of a program's reduct P^X via operator T_{P^X} nicely illustrates the constructive nature of stable models. Given that stable models equal $Cn(P^X)$, each of their atoms must be derived by successively applying rules (and ultimately facts) in P^X. This is made precise in Section 5.1.

For computing stable models of normal programs, we decompose their fixpoint characterization for developing an approximation scheme. To this end, we approximate a stable model X by two sets of atoms L and U such that $L \subseteq X \subseteq U$. That is, L and U constitute lower and upper bounds on X, respectively. One may view L and $(\mathcal{A} \setminus U)$ as the true and false atoms in a partial three-valued interpretation of the program, respectively. Now, for sets X and Y of ground atoms and a logic program P, a key observation is that

$$X \subseteq Y \quad \text{implies} \quad P^Y \subseteq P^X \quad \text{implies} \quad Cn(P^Y) \subseteq Cn(P^X).$$

Given a stable model X of P, this entails the following properties:

- If $L \subseteq X$, then $X \subseteq Cn(P^L)$.

- If $X \subseteq U$, then $Cn(P^U) \subseteq X$.

- If $L \subseteq X \subseteq U$, then $L \cup Cn(P^U) \subseteq X \subseteq U \cap Cn(P^L)$.

The last property provides a recipe for iteratively tightening the lower and upper bounds:

repeat

    replace L with $L \cup Cn(P^U)$
    replace U with $U \cap Cn(P^L)$

until L and U do not change anymore.

We observe that L becomes larger (or remains the same) and U becomes smaller (or remains the same) at each iteration. Nonetheless, the property $L \subseteq X \subseteq U$ is invariant for each stable model X of P. Thus, the above recipe tightens the approximation while preserving all enclosed stable models. If we get $L \not\subseteq U$ at some point, the original bounds encompassed no stable model. However, once we obtain $L = U$, then L is a stable model of P.

For illustration, consider Program P_7 below.

$$P_7 \; = \; \left\{ \begin{array}{lll} a \leftarrow & c \leftarrow a, {\sim}d & e \leftarrow b, {\sim}f \\ b \leftarrow {\sim}a & d \leftarrow {\sim}c, {\sim}e & e \leftarrow e \end{array} \right\}$$

Starting with $L_0 = \emptyset$ and $U_0 = atom(P_7) = \{a, b, c, d, e, f\}$ results in the following.[6]

$$P_7^{\emptyset} \; = \; \left\{ \begin{array}{lll} a \leftarrow & c \leftarrow a & e \leftarrow b \\ b \leftarrow & d \leftarrow & e \leftarrow e \end{array} \right\} \qquad P_7^{\{a,b,c,d,e,f\}} \; = \; \left\{ \begin{array}{ll} a \leftarrow & \\ & e \leftarrow e \end{array} \right\}$$

$$Cn(P_7^{\emptyset}) \; = \; \{a, b, c, d, e\} \qquad\qquad\qquad\qquad Cn(P_7^{\{a,b,c,d,e,f\}}) \; = \; \{a\}$$

Consequently, we get $L_1 = \{a\}$ and $U_1 = \{a, b, c, d, e\}$, showing that a belongs to all and f to no stable models of P_7, respectively. The new bounds yield the following reduced programs and their sets of consequences.

$$P_7^{\{a\}} \; = \; \left\{ \begin{array}{lll} a \leftarrow & c \leftarrow a & e \leftarrow b \\ & d \leftarrow & e \leftarrow e \end{array} \right\} \qquad P_7^{\{a,b,c,d,e\}} \; = \; \left\{ \begin{array}{ll} a \leftarrow & e \leftarrow b \\ & e \leftarrow e \end{array} \right\}$$

$$Cn(P_7^{\{a\}}) \; = \; \{a, c, d\} \qquad\qquad\qquad\qquad Cn(P_7^{\{a,b,c,d,e\}}) \; = \; \{a\}$$

This results in $L_2 = \{a\}$ and $U_2 = \{a, c, d\}$. The latter reflects that in addition to f neither b nor e belong to any stable model of P_7. We get the following reducts along with their smallest models.

$$P_7^{\{a\}} \; = \; \left\{ \begin{array}{lll} a \leftarrow & c \leftarrow a & e \leftarrow b \\ & d \leftarrow & e \leftarrow e \end{array} \right\} \qquad P_7^{\{a,c,d\}} \; = \; \left\{ \begin{array}{ll} a \leftarrow & e \leftarrow b \\ & e \leftarrow e \end{array} \right\}$$

$$Cn(P_7^{\{a\}}) \; = \; \{a, c, d\} \qquad\qquad\qquad\qquad Cn(P_7^{\{a,c,d\}}) \; = \; \{a\}$$

We finally reached a fixpoint resulting in $L_3 = \{a\}$ and $U_3 = \{a, c, d\}$. These bounds correspond to the partial interpretation $(\{a\}, \{b, e, f\})$, making explicit which atoms belong to each and no stable model of P_7, respectively.

The above recipe allows us to compute the (deterministic) consequences of a program relative to some given bounds. In other words, it provides a blueprint for designing a propagation algorithm. To this end, let $expand_P(L, U)$ denote the result obtained by computing a fixpoint with the above recipe starting from bounds L and U. In our example, we thus have $expand_{P_7}(\emptyset, atom(P_7)) = (\{a\}, \{a, c, d\})$. Given that each stable model X of P_7 (trivially) satisfies the condition $\emptyset \subseteq X \subseteq atom(P_7)$, the aforementioned preservation of stable models guaranteed by $expand_P$ implies also the

---

[6]We index L and U to indicate the respective iteration step.

strengthened condition $\{a\} \subseteq X \subseteq \{a, c, d\}$. This leaves us with four candidate models, depending on the membership of c and d, respectively.

The idea is now to proceed by case analysis. To this end, we choose either c or d and consider in turn all candidate sets including and excluding the chosen atom. For example, let us decide to (non-deterministically) choose c. Then, we first explore model candidates containing c by adding it to the lower bound. Proceeding analogously as above, we obtain

- $expand_{P_7}(\{a\} \cup \{c\}, \{a, c, d\}) = (\{a, c\}, \{a, c\})$

- $expand_{P_7}(\{a\}, \{a, c, d\} \setminus \{c\}) = (\{a, d\}, \{a, d\})$

In both cases, we get a stable model of P_7, yielding $\{a, c\}$ and $\{a, d\}$, respectively. These two-valued models correspond to the total three-valued interpretations $(\{a, c\}, \{b, d, e, f\})$ and $(\{a, d\}, \{b, c, e, f\})$, respectively. In both computations, a, b, d, e, and f constitute deterministic consequences because their truth assignment was computed within $expand_p$, whereas the respective truth assignment to c is regarded as a non-deterministic consequence because c was chosen arbitrarily. The computation of deterministic consequences is also referred to as *propagation*. The process of deciding upon non-deterministic consequences is simply called *choice*.

The successive alternation of propagation and choice constitutes a general computation scheme in Constraint Programming. Building upon the above approximation strategy encapsulated in $expand_p$, we can instantiate this general scheme in the following search algorithm. As above, let $L \subseteq U \subseteq atom(P)$ for some logic program P.

$solve_P(L, U)$

$(L, U) \leftarrow expand_p(L, U)$ // propagation

if $L \not\subseteq U$ **then failure** // failure

if $L = U$ **then output** L // success

else choose $a \in U \setminus L$ // choice

$\qquad solve_P(L \cup \{a\}, U)$

$\qquad solve_P(L, U \setminus \{a\})$

The call $solve_P(\emptyset, atom(P))$ leads to outputting all stable models of P. We observe that the above algorithm traverses a search tree spanned by the chosen atoms and the following case analysis. This amounts to a time complexity of $O(2^{|atom(P)|})$. The actual number of choices among $atom(P)$ is influenced by the strength of the propagation and the quality of the choices because one choice may result in more deterministic consequences than another.

2.4.2 REASONING MODES

Although determining whether a program has a stable model is the fundamental decision problem in ASP, more reasoning modes are needed for covering the variety of reasoning problems encountered in applications. Also, in practice, one is often less interested in solving decision problems but rather in solving the corresponding search problems.

The most prominent problems solved by ASP solvers are the following. To capture these, let $SM(P)$ denote the set of stable models of a given program P.

Satisfiability Compute some $X \in SM(P)$
   (implicitly deciding whether a stable model exists).

Unsatisfiability Show that $SM(P) = \emptyset$.

Enumeration Enumerate (n elements of) $SM(P)$.

Projection Enumerate (n elements of) $\{X \cap \mathcal{B} \mid X \in SM(P)\}$ for $\mathcal{B} \subseteq \mathcal{A}$.

Intersection Compute $\bigcap_{X \in SM(P)} X$.

Union Compute $\bigcup_{X \in SM(P)} X$.

Optimization Compute some (or: Enumerate n elements of) $X \in \arg\min_{X \in SM(P)} \nu(X)$

   where ν is an objective function subject to minimization, as expressed by optimization statements in P.

2.4.3   COMPUTATIONAL COMPLEXITY

This section briefly summarizes complexity results for the basic decision problems in ASP.
   In what follows, let a be an atom and X be a set of atoms.

* For a positive normal logic program P:

   – Deciding whether X is the stable model of P is P-complete.
   – Deciding whether a is in the stable model of P is P-complete.

* For a normal logic program P:

   – Deciding whether X is a stable model of P is P-complete.
   – Deciding whether a is in a stable model of P is NP-complete.

* For a normal logic program P with optimization statements:

   – Deciding whether X is an optimal stable model of P is *co-NP*-complete.
   – Deciding whether a is in an optimal stable model of P is Δ_2^P-complete.

* For a positive disjunctive logic program P:

   – Deciding whether X is a stable model of P is *co-NP*-complete.

  – Deciding whether a is in a stable model of P is NP^{NP}-complete.

- For a disjunctive logic program P:

  – Deciding whether X is a stable model of P is *co-NP*-complete.

  – Deciding whether a is in a stable model of P is NP^{NP}-complete.

- For a disjunctive logic program P with optimization statements:

  – Deciding whether X is an optimal stable model of P is *co-NP^{NP}*-complete.

  – Deciding whether a is in an optimal stable model of P is Δ_3^P-complete.

- For a propositional theory Φ:

  – Deciding whether X is a stable model of Φ is *co-NP*-complete.

  – Deciding whether a is in a stable model of Φ is NP^{NP}-complete.

The above complexity results apply to propositional programs only. For capturing the complexity of the first-order case, we note that the ground instantiation $grd(P)$ of a first-order program P is exponential in the size of P. Hence, roughly speaking, we obtain the analogous results by replacing the base *NP* by *NEXPTIME* for capturing programs with variables.

2.5   REFERENCES AND FURTHER READING

Introductory textbooks on Logic and Logic Programming can be found in (Enderton, 1972, Gallier, 1986, Lloyd, 1987). A basic tutorial on Logic Programming leading to ASP can be found in (Lifschitz, 2004).

A positive rule is also referred to as being definite; it corresponds to a disjunction with exactly one positive literal, which itself is called a definite clause. Horn clauses are clauses with at most one positive atom. Clearly, each definite clause is a Horn clause but not vice versa. Informally, Horn clauses can be seen as an extension of definite clauses with integrity constraints. A set of definite (Horn) clauses has a $\subseteq$-smallest model (or none). Given a positive program P, its set of consequences $Cn(P)$ corresponds to the smallest model of the set of definite clauses corresponding to P (see also the definition of $\overleftarrow{CF}(P)$ on Page 69). For details, we refer the reader to Lifschitz (2004), Lloyd (1987).

The *stable models semantics* was defined by Michael Gelfond and Vladimir Lifschitz in (1988), and further elaborated in (1991). The term *answer set* was introduced in Gelfond and Lifschitz (1990), when extending the stable models semantics to logic programs with classical negation. Such programs admit "stable models" containing both or neither of an atom and its classical negation; hence they were referred to as answer sets.

An alternative reduct was defined by Faber et al. (2004); it amounts to the one given in Linke (2001), which was itself derived from the characterization of "default extensions" through "generating rules" in Default Logic (Reiter, 1980).

The development of the algorithms *expand*$_P$ and *solve*$_P$ from the definition of stable models is due to Lifschitz (2004). The algorithmic scheme of *solve*$_P$ follows that of the well-known Davis-Putman-Logemann-Loveland (DPLL) procedure (Davis and Putnam, 1960, Davis et al., 1962) for SAT. The first implementation computing the stable models of a logic program is called *smodels* and due to Niemelä and Simons (1995), Simons et al. (2002); it is detailed in the dissertation of Simons (2000). Our naming of the above propagation procedure credits the one in *smodels*, which is called *expand*. Interestingly, *expand* is composed of two subprocedures *atleast* and *atmost*. While *atleast* corresponds to the (Fitting) operator Φ_P described in Section 5.2, *atmost* computes the greatest unfounded set, referred to as $\mathbf{U}_P$ in Section 5.2.

Even and odd cycles in networks constitute an omnipresent pattern in informatics (and natural sciences). Examples of this in ASP are given in Program P_2 and P_3. While the two rules in P_2 form an even cycle (in the underlying atom dependency graph) inducing two alternative solutions, the single rule in P_3 induces an odd cycle denying a solution.

Viewing rules with first-order variables as schemas representing their sets of ground instances is common in Logic Programming (Lloyd, 1987). The semantic counterpart of this is provided by concepts developed by Jacques Herbrand. In view of this, the set of terms and atoms are also called Herbrand universe and Herbrand base, respectively. A first-order interpretation whose underlying domain is the Herbrand universe is called a Herbrand interpretation. Each Herbrand interpretation corresponds to a subset of the Herbrand base. See (Lloyd, 1987) for a formal account of these concepts.

A theory on least and greatest fixpoints for monotone operators on lattices is given in Tarski (1955). The Knaster-Tarski Theorem tells us that each monotone operator on a complete lattice has a least fixpoint. See Eiter et al. (2009), Lifschitz (2004) for brief introductions.

The material of Section 2.3.2 is largely based on the work of Simons et al. comprehensively published in (2002). In fact, choice and cardinality rules as well as cardinality and weight constraints were introduced with the *lparse/smodels* systems (cf. Niemelä and Simons (1997), Syrjänen). The same applies to optimization statements. The semantics of choice, cardinality, and weight rules is defined from first principles in Simons et al. (2002), as well as basic translations from which we derived the ones in Section 2.3.

Weak constraints and tuple-oriented forms of cardinality and weight constraints are due to Leone et al. (2006), and implemented originally within the ASP system *dlv*. They are also available in (the experimental) *gringo* version 3.0.92 and will be included in the upcoming *gringo* series 4.

Explicit priority levels are supported in recent versions of the grounder *gringo* (Gebser et al.). This avoids a dependency of priorities on input order, which is considered by *lparse* (Syrjänen) if several minimize statements are provided. Priority levels are also supported by *dlv* (Leone et al., 2006) in weak constraints. Furthermore, we admit negative weights in minimize statements, where they cannot raise semantic problems (cf. Ferraris (2005)) going along with the rewriting of weight constraints suggested in Simons et al. (2002).

Conditional literals were introduced with the grounder *lparse* (Syrjänen); see Syrjänen (2004) for a formal introduction.

The elimination of default negation in the head is due to Janhunen (2001).

Many other forms of preferences for specifying optimal stable models can be found in the literature. An overview is given in Delgrande et al. (2004). Gebser et al. (2011h) show how minimize (and maximize) statements can be used for expressing inclusion and Pareto-based preference criteria.

Disjunctive logic programs under stable models semantics were introduced in Gelfond and Lifschitz (1991). The addition of default negation to disjunctive heads is due to Lifschitz and Woo (1992).

The stable models semantics for propositional theories was defined in Ferraris (2005). Cabalar and Ferraris (2007) show that propositional theories are (strongly) equivalent to disjunctive logic programs.

Dantsin et al. (1997) give a survey of complexity results in logic programming. A fine-grained account of subclasses of programs is given in Truszczynski (2011). See also Simons et al. (2002) and Leone et al. (2006) for details on computational complexity. The lifting of complexity results from the ground to the non-ground case is described in Leone et al. (2006).

CHAPTER 3

Basic modeling

This chapter uses three well-known examples to illustrate some basic principles of modeling in ASP. First, we use graph coloring to show how problem classes and instances are encoded as logic programs. Then, we stepwise develop a solution to the n-queens problem for illustrating ASP's modeling methodology along with the workflow underlying ASP's solving process. Finally, we deal with the traveling salesperson problem in order to provide some interesting modeling and solving details.

3.1 PROBLEM ENCODING

For solving a problem class **C** for a problem instance **I** in ASP, we encode

1. the problem instance **I** as a set P_I of facts and

2. the problem class **C** as a set P_C of rules

such that the solutions to **C** for **I** can be (polynomially) extracted from the stable models of $P_I \cup P_P$. An encoding P_C is *uniform*, if it can be used to solve all its problem instances. That is, P_C encodes the solutions to **C** for any set P_I of facts. Note that propositional logic admits no uniform encodings in general; rather each problem instance needs to be encoded separately.

As a first illustrative example, let us consider graph coloring. More precisely, given a graph (V, E), assign each node in V one of n colors such that no two nodes in V connected by an edge in E have the same color. A problem instance is given by the graph, whereas the problem class consists in finding assignments of colors to nodes subject to the given constraint. As an example, consider Graph G_8 depicted in Figure 3.1.

$$G_8 = \left(\left\{ \begin{array}{c} 1 \\ 2 \\ 3 \\ 4 \\ 5 \\ 6 \end{array} \right\}, \left\{ \begin{array}{ccc} (1,2) & (1,3) & (1,4) \\ (2,4) & (2,5) & (2,6) \\ (3,1) & (3,4) & (3,5) \\ (4,1) & (4,2) & \\ (5,3) & (5,4) & (5,6) \\ (6,2) & (6,3) & (6,5) \end{array} \right\} \right) \qquad (3.1)$$

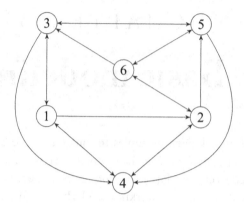

Figure 3.1: Directed graph G_8 having 6 nodes and 17 edges.

This graph is represented by predicates node/1 and edge/2 in Program P_8 given in Listing 3.1.

Listing 3.1: Program P_8 representing Graph 3.1 (graph.lp)

```
1  node(1..6).

3  edge(1,2).   edge(1,3).   edge(1,4).
4  edge(2,4).   edge(2,5).   edge(2,6).
5  edge(3,1).   edge(3,4).   edge(3,5).
6  edge(4,1).   edge(4,2).
7  edge(5,3).   edge(5,4).   edge(5,6).
8  edge(6,2).   edge(6,3).   edge(6,5).
```

The two dots .. in node(1..6) allow us to represent the six instances of node/1 in a compact way (see below).

The 3-colorability problem is encoded in Program P_9 given in Listing 3.2.

Listing 3.2: Program P_9 encoding graph coloring (color.lp)

```
1  col(r). col(g). col(b).

3  1 {color(X,C) : col(C)} 1 :- node(X).
4  :- edge(X,Y), color(X,C), color(Y,C).

6  #hide.
7  #show color/2.
```

Line 1 provides the available colors. (Arguably this information could also be included in the problem instance.) Lines 3 and 4 encode the actual colorability problem. The first rule expresses that each node X must be colored with exactly one color among red, blue, and green. Without the integrity constraint in Line 4 this would result in all possible colorings, including those painting a whole graph with the same color. Such colorings are however ruled out in Line 4 by stipulating that connected nodes must be colored differently. Finally, Lines 6 and 7 are directives advising the solver to project stable models onto instances of predicate color/2.

Looking at the rules in Line 3 and 4, we note that each variable is bound via an occurrence in a positive literal. Among them, we distinguish local and global variables. For instance, variable C in Line 3 is local to the encompassing cardinality constraint; it is bound by col(C) and varies over all instantiations of col(C). All other variables are global with respect to each rule and must be bound by a positive body literal. For instance, X acts as a global variable in Line 3; it is bound through the positive body literal node(X). Accordingly, one rule instance is produced for each instantiation of node(X) (unless simplifications apply). Rules satisfying these requirements are said to be *safe*, as detailed in Chapter 4.

Following the ASP solving process in Figure 1.2 on Page 3, let us now instantiate the combined Program $P_8 \cup P_9$. This first processing step is displayed in Listing 3.3. [1]

Listing 3.3: Grounding Program $P_8 \cup P_9$

```
$ gringo --text color.lp graph.lp

col(r). col(g). col(b).

edge(1,2). edge(1,3). edge(1,4). edge(2,4). edge(2,5). edge(2,6).
edge(3,1). edge(3,4). edge(3,5). edge(4,1). edge(4,2). edge(5,3).
edge(5,4). edge(5,6). edge(6,2). edge(6,3). edge(6,5).

node(1). node(2). node(3). node(4). node(5). node(6).

1 #count {color(6,b),color(6,g),color(6,r)} 1.
1 #count {color(5,b),color(5,g),color(5,r)} 1.
1 #count {color(4,b),color(4,g),color(4,r)} 1.
1 #count {color(3,b),color(3,g),color(3,r)} 1.
1 #count {color(2,b),color(2,g),color(2,r)} 1.
1 #count {color(1,b),color(1,g),color(1,r)} 1.

:-color(1,b),color(2,b).  :-color(2,b),color(4,b).  :-color(3,b),color(1,b).
:-color(1,b),color(3,b).  :-color(2,b),color(5,b).  :-color(3,b),color(4,b).
:-color(1,b),color(4,b).  :-color(2,b),color(6,b).  :-color(3,b),color(5,b).
:-color(1,g),color(2,g).  :-color(2,g),color(4,g).  :-color(3,g),color(1,g).
:-color(1,g),color(3,g).  :-color(2,g),color(5,g).  :-color(3,g),color(4,g).
:-color(1,g),color(4,g).  :-color(2,g),color(6,g).  :-color(3,g),color(5,g).
:-color(1,r),color(2,r).  :-color(2,r),color(4,r).  :-color(3,r),color(1,r).
:-color(1,r),color(3,r).  :-color(2,r),color(5,r).  :-color(3,r),color(4,r).
:-color(1,r),color(4,r).  :-color(2,r),color(6,r).  :-color(3,r),color(5,r).

:-color(4,b),color(1,b).  :-color(5,b),color(3,b).  :-color(6,b),color(2,b).
:-color(4,b),color(2,b).  :-color(5,b),color(4,b).  :-color(6,b),color(3,b).
:-color(4,g),color(1,g).  :-color(5,b),color(6,b).  :-color(6,b),color(5,b).
```

[1]The resulting output has been reformatted for the sake of conciseness.

```
:-color(4,g),color(2,g).  :-color(5,g),color(3,g).  :-color(6,g),color(2,g).
:-color(4,r),color(1,r).  :-color(5,g),color(4,g).  :-color(6,g),color(3,g).
:-color(4,r),color(2,r).  :-color(5,g),color(6,g).  :-color(6,g),color(5,g).
                          :-color(5,r),color(3,r).  :-color(6,r),color(2,r).
                          :-color(5,r),color(4,r).  :-color(6,r),color(3,r).
                          :-color(5,r),color(6,r).  :-color(6,r),color(5,r).

#hide.
#show color/2.
```

Line 1 shows the command prompt, $, followed by the command launching the grounder *gringo* on the files color.lp and graph.lp containing programs P_9 and P_8, respectively. The option --text tells *gringo* to print the ground program in a human readable way. We observe that the grounder proceeds in a "bottom-up" fashion by first outputting all factual information. The following six ground cardinality constraints result from the ground instantiation of Rule 3 in Listing 3.2. One such constraint is generated for each node. The respective instance of the node/1 predicate in the original body was evaluated and subsequently eliminated. Also, the grounder made the fact explicit that cardinality constraints are count aggregates, indicated by #count. Similarly, ground integrity constraints are obtained from the integrity constraint in Line 4 of Listing 3.2. We obtain 51 ground rules in view of 17 edges and 3 colors. Again, true parts, viz. instances of edge/2, have been simplified away. The last two lines just repeat the solver directives from Listing 3.2.

For computing the stable models of the resulting ground program, it is passed in machine readable format (by omitting the *gringo* option --text) to the ASP solver *clasp*.

Listing 3.4: Grounding and solving Program $P_8 \cup P_9$

```
$ gringo color.lp graph.lp | clasp 0
clasp version 2.0.5
Reading from stdin
Solving...
Answer: 1
color(6,g) color(5,b) color(4,g) color(3,r) color(2,r) color(1,b)
Answer: 2
color(6,r) color(5,b) color(4,r) color(3,g) color(2,g) color(1,b)
Answer: 3
color(6,b) color(5,g) color(4,b) color(3,r) color(2,r) color(1,g)
Answer: 4
color(6,r) color(5,g) color(4,r) color(3,b) color(2,b) color(1,g)
Answer: 5
color(6,b) color(5,r) color(4,b) color(3,g) color(2,g) color(1,r)
Answer: 6
color(6,g) color(5,r) color(4,g) color(3,b) color(2,b) color(1,r)
SATISFIABLE

Models     : 6
Time       : 0.000s (Solving: 0.00s 1st Model: 0.00s Unsat: 0.00s)
CPU Time   : 0.000s
```

The option 0 instructs *clasp* to compute all stable models. In our example, we obtain six of them, each representing a different coloring of Graph G_8 in Figure 3.1. The coloring of G_8 expressed by the fifth stable model is given in Figure 3.2.

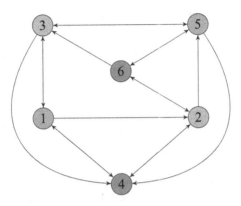

Figure 3.2: A coloring of Graph G_8.

3.2 MODELING METHODOLOGY

The basic approach to writing encodings in ASP follows a *generate-and-test* methodology, also referred to as *guess-and-check*, inspired by intuitions on *NP* problems. A "generating" part is meant to non-deterministically provide solution candidates, while a "testing" part eliminates candidates violating some requirements. (Note that this decomposition is only a methodological one; it is neither syntactically enforced nor computationally relevant.) Both parts are usually amended by "defining" parts providing auxiliary concepts. In addition, one may specify optimization criteria via lexicographically ordered objective functions, as described in Section 2.3.3.

In fact, the coloring encoding in Listing 3.2 is an exemplar for this methodology. The rule in Line 3 generates solution candidates, among which the integrity constraint in Line 4 tests whether they constitute valid solutions.

For further illustrating this methodology, let us stepwisely develop a solution to the n-queens problem. For this, we want to place n queens on an $n \times n$ chess board such that no two queens attack each other.

To begin with, we define in Listing 3.5 the underlying board consisting of n rows and n columns.

Listing 3.5: Program P_{10} addressing the n-queens problem, Part I (queensI.lp)

```
1  row(1..n).
2  col(1..n).
```

This simple program has a single stable model (for each given value of n), as shown in Listing 3.6.

Listing 3.6: Grounding and solving Program P_{10} (for n=5)

```
$ gringo queens0.lp --const n=5 | clasp 0
clasp version 2.0.5
Reading from stdin
Solving...
Answer: 1
row(1) row(2) row(3) row(4) row(5) col(1) col(2) col(3) col(4) col(5)
SATISFIABLE

Models    : 1
Time      : 0.000s (Solving: 0.00s 1st Model: 0.00s Unsat: 0.00s)
CPU Time  : 0.000s
```

Note that the option '--const n=5' tells the grounder to replace the constant n in Program P_{10} by 5. This gives our representation of a 5 × 5-chess board, as illustrated in Figure 3.3. The argument 0 makes the solver enumerate all stable models, although there is only one, as indicated by its output 'Models : 1'.

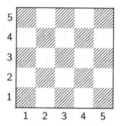

Figure 3.3: Board corresponding to the stable model of Logic program P_{10}

Next, let us populate the board in Figure 3.3. In view of our methodology, we generate solution candidates by placing queens on the board. This is accomplished by the choice rule in Line 4 of Listing 3.7. From now on, we project stable models on instances of the queen/2 predicate, as indicated in Line 6.

Listing 3.7: Program P_{11} addressing the n-queens problem, Part II (queensII.lp)

```
1  row(1..n).
2  col(1..n).

4  { queen(I,J) : col(I) : row(J) }.

6  #hide.  #show queen/2.
```

Program P_{11} has 33554432 stable models. For brevity, let us instruct the solver *clasp* to enumerate only five of them. The result is shown in Listing 3.8. The obtained stable models are illustrated in Figure 3.4.

Listing 3.8: Grounding and solving Program P_{11}

```
$ gringo queens1.lp --const n=5 | clasp 5
clasp version 2.0.5
Reading from stdin
Solving...
Answer: 1

Answer: 2
queen(1,1)
Answer: 3
queen(2,1)
Answer: 4
queen(2,1)  queen(1,1)
Answer: 5
queen(3,1)
SATISFIABLE

Models      : 5+
Time        : 0.000s  (Solving: 0.00s 1st Model: 0.00s Unsat: 0.00s)
CPU Time    : 0.000s
```

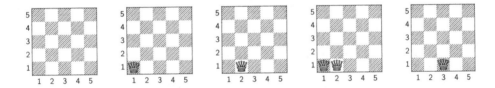

Figure 3.4: Boards corresponding to five stable models of Logic program P_{11}

Unlike Listing 3.6, where *clasp* indicates that there is exactly one stable model of P_{10}, it tells us via 5+ that more than five stable models of P_{11} could be enumerated. In fact, the enumeration of all models would at some point also yield a chessboard full of queens. However, most of these solution candidates are invalid. In fact, all five examples fail to place the necessary number of queens.

To this end, we add an integrity constraint eliminating all boards with either fewer or more than n queens. The resulting program, P_{12}, is given in Listing 3.9.

Listing 3.9: Program P_{12} addressing the n-queens problem, Part III (queensIII.lp)

```
1  row(1..n).
2  col(1..n).
```

```
4  { queen(I,J) : col(I) : row(J) }.
5   :- not n { queen(I,J) } n.

7  #hide.  #show queen/2.
```

Actually, the same effect is obtained by replacing Lines 4 and 5 of Listing 3.9 with

```
    n { queen(I,J) : row(I) : col(J) } n.
```

We simply opted for the modular change in view of a successive development of our example.

In fact, Line 5 reveals a particularity of *gringo*'s grounding procedure. Strictly speaking, the instantiation of variables I and J is unrestricted within the cardinality constraint. However, this is tolerated by *gringo* because the instantiation of queen/2 is delineated in Line 4. Hence, the bound established in Line 4 is also applied to the instantiation of I and J in queen(I,J) in Line 5 (unless further restrictions are given).

Although Program P_{12} eliminates many invalid solutions obtained from P_{11}, it still possesses 53130 stable models. In analogy to the above, we compute five of them in Listing 3.10 and illustrate them in Figure 3.5.

Listing 3.10: Grounding and solving Program P_{12}

```
$ gringo queens2.lp --const n=5 | clasp 5
clasp version 2.0.5
Reading from stdin
Solving...
Answer: 1
queen(5,1) queen(4,1) queen(3,1) queen(2,1) queen(1,1)
Answer: 2
queen(1,2) queen(4,1) queen(3,1) queen(2,1) queen(1,1)
Answer: 3
queen(1,2) queen(5,1) queen(3,1) queen(2,1) queen(1,1)
Answer: 4
queen(1,2) queen(5,1) queen(4,1) queen(2,1) queen(1,1)
Answer: 5
queen(1,2) queen(5,1) queen(4,1) queen(3,1) queen(1,1)
SATISFIABLE

Models    : 5+
Time      : 0.000s (Solving: 0.00s 1st Model: 0.00s Unsat: 0.00s)
CPU Time  : 0.000s
```

We observe that all five example solutions have multiple queens in the same line and/or row. For eliminating such solution candidates, we add two integrity constraints forbidding horizontal and vertical attacks. The resulting program, P_{13}, is given in Listing 3.11.

Listing 3.11: Program P_{13} addressing the *n*-queens problem, Part IV (queensIV.lp)

```
1  row(1..n).
```

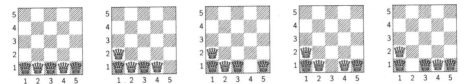

Figure 3.5: Boards corresponding to five stable models of Logic program P_{12}.

```
2  col(1..n).

4  { queen(I,J) : col(I) : row(J) }.
5    :- not n { queen(I,J) } n.
6    :- queen(I,J), queen(I,JJ), J != JJ.
7    :- queen(I,J), queen(II,J), I != II.

9  #hide.  #show queen/2.
```

In total, we obtain 120 stable models from P_{13}, among which we compute and illustrate again five examples in Listing 3.12 and Figure 3.6, respectively.

Listing 3.12: Grounding and solving Program P_{13}

```
$ gringo queens3.lp --const n=5 | clasp 5
clasp version 2.0.5
Reading from stdin
Solving...
Answer: 1
queen(5,5) queen(4,4) queen(3,3) queen(2,2) queen(1,1)
Answer: 2
queen(5,5) queen(4,4) queen(3,3) queen(1,2) queen(2,1)
Answer: 3
queen(5,5) queen(4,4) queen(2,3) queen(1,2) queen(3,1)
Answer: 4
queen(5,5) queen(4,4) queen(1,3) queen(2,2) queen(3,1)
Answer: 5
queen(5,5) queen(4,4) queen(1,3) queen(3,2) queen(2,1)
SATISFIABLE

Models     : 5+
Time       : 0.000s (Solving: 0.00s 1st Model: 0.00s Unsat: 0.00s)
CPU Time   : 0.000s
```

We still observe that the five selected solution candidates are invalid because of multiple queens in diagonals. As above, we add two integrity constraints forbidding diagonal attacks. This yields Program P_{14} given in Listing 3.13.

Listing 3.13: Program P_{14} solving the *n*-queens problem (queens.lp)

Figure 3.6: Boards corresponding to five stable models of Logic program P$_{13}$.

```
1  row(1..n).
2  col(1..n).

4  { queen(I,J) : col(I) : row(J) }.
5  :- not n { queen(I,J) } n.
6  :- queen(I,J), queen(I,JJ), J != JJ.
7  :- queen(I,J), queen(II,J), I != II.
8  :- queen(I,J), queen(II,JJ), (I,J) != (II,JJ), I-J == II-JJ.
9  :- queen(I,J), queen(II,JJ), (I,J) != (II,JJ), I+J == II+JJ.

11 #hide.  #show queen/2.
```

In fact, Lines 6 to 9 contain arithmetic functions + and − as well as comparison predicates != and ==. (To guarantee their safe instantiation, each contained variable must be bound by a positive body literal.) Once instantiated, the corresponding expressions are evaluated by the grounder and resulting simplifications are applied. For instance, no ground rule is produced from Line 6 whenever J and JJ are instantiated with the same term; otherwise the inequality holds and is removed from the resulting ground rule.

Finally, Program P$_{14}$ has ten stable models, all of which are now solutions to the 5-queens problem. The computation is given in Listing 3.14, and the resulting solutions are illustrated in Figure 3.7.

Listing 3.14: Grounding and solving Program P$_{14}$

```
$ gringo queens.lp --const n=5 | clasp 0
clasp version 2.0.5
Reading from stdin
Solving...
Answer: 1
queen(4,5) queen(1,4) queen(3,3) queen(5,2) queen(2,1)
Answer: 2
queen(2,5) queen(5,4) queen(3,3) queen(1,2) queen(4,1)
Answer: 3
queen(2,5) queen(4,4) queen(1,3) queen(3,2) queen(5,1)
Answer: 4
queen(4,5) queen(2,4) queen(5,3) queen(3,2) queen(1,1)
Answer: 5
```

```
queen(5,5) queen(2,4) queen(4,3) queen(1,2) queen(3,1)
Answer: 6
queen(1,5) queen(4,4) queen(2,3) queen(5,2) queen(3,1)
Answer: 7
queen(5,5) queen(3,4) queen(1,3) queen(4,2) queen(2,1)
Answer: 8
queen(1,5) queen(3,4) queen(5,3) queen(2,2) queen(4,1)
Answer: 9
queen(3,5) queen(1,4) queen(4,3) queen(2,2) queen(5,1)
Answer: 10
queen(3,5) queen(5,4) queen(2,3) queen(4,2) queen(1,1)
SATISFIABLE

Models      : 10
Time        : 0.001s (Solving: 0.00s 1st Model: 0.00s Unsat: 0.00s)
CPU Time    : 0.000s
```

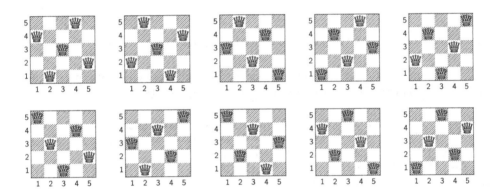

Figure 3.7: Boards corresponding to the stable models of Logic program P_{14}

All in all, we have seen how the successive addition of integrity constraints narrows down the solution candidates. Apart from illustrating the *generate-and-test* methodology of ASP, it shows its elaboration tolerance.

3.3   ADVANCED PROBLEM ENCODING

Finally, we consider the well-known traveling salesperson problem. The task is to decide whether there is a round trip visiting each node in a graph exactly once (also known as a Hamiltonian cycle) such that accumulated edge costs do not exceed some budget. We tackle a slightly more general variant of the problem by not a priori fixing the budget. Rather, we want to compute a round trip with a minimum budget.

For this purpose, let us reconsider Graph G_8 but associate costs with edges. Figure 3.8 shows the augmented graph from Figure 3.1. Symmetric edges have the same costs here, but differing costs would be possible as well.

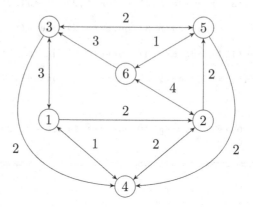

Figure 3.8: Graph G_8 from Figure 3.1 annotated with edge costs.

To accommodate edge costs, we augment the graph representation expressed in Program P_8 (cf. Listing 3.1) by the facts in Program P_{15} given in Listing 3.15.

Listing 3.15: Program P_{15} representing edge costs of Graph G_8 (costs.lp)

```
1   cost (1,2,2).    cost (1,3,3).    cost (1,4,1).
2   cost (2,4,2).    cost (2,5,2).    cost (2,6,4).
3   cost (3,1,3).    cost (3,4,2).    cost (3,5,2).
4   cost (4,1,1).    cost (4,2,2).
5   cost (5,3,2).    cost (5,4,2).    cost (5,6,1).
6   cost (6,2,4).    cost (6,3,3).    cost (6,5,1).
```

Program P_{15} contains an instance of cost/3 for each (directed) edge in Figure 3.8.

As mentioned above, the first subproblem consists of finding a round trip, giving a candidate for a minimum-cost round trip. Following our *generate-and-test* methodology, we encode this subproblem via the following rules.

Listing 3.16: Program P_{16} addressing the round trip (Hamiltonian cycle) problem (ham.lp)

```
1   1 { cycle(X,Y) : edge(X,Y) } 1 :- node(X).
2   1 { cycle(X,Y) : edge(X,Y) } 1 :- node(Y).

4   reached(Y) :- cycle(1,Y).
5   reached(Y) :- cycle(X,Y), reached(X).
```

```
7   :- node(Y), not reached(Y).

9   #hide.   #show cycle/2.
```

Solution candidates are represented by instances of predicate `cycle/2`, chosen among all edges of Graph G_8. Accordingly, Line 9 projects stable models on the respective instances.

The "generating" rules in Lines 1 and 2 make sure that each node in a graph must have exactly one outgoing and exactly one incoming edge, respectively. These edges are captured by means of predicate `cycle/2`. Let us make this more precise by inserting the available edges for node 1. This yields the following instantiation of Lines 1 and 2:

```
1 { cycle(1,2), cycle(1,3), cycle(1,4) } 1.
1 { cycle(3,1), cycle(4,1) } 1.
```

The first rule groups all outgoing edges of node 1, while the second does the same for its incoming edges. Together both rules provide us with six possibilities to get across node 1.

The two rules in Lines 4 and 5 are "defining" rules, which (recursively) determine which nodes are reached by a cycle candidate produced in the "generating" part. Note that the rule in Line 4 builds on the assumption that the cycle "starts" at node 1, that is, any successor Y of 1 is reached by the cycle. The second rule in Line 5 states that, from a reached node X, an adjacent node Y can be reached via a further edge in the cycle. Notably, the definition of `reached/1` in Lines 4 and 5 relies on an adequate treatment of positive recursion (among ground instances of `reached/1`).[2] That is, only derivable atoms are made true, while all others are set to false. This feature makes sure that all nodes are reached by a global cycle from node 1, thus, excluding isolated subcycles. In fact, the "test" in Line 7 ensures that each node in a given graph is reached, that is, the instances of `cycle/2` in a stable model must be edges of a round trip.

Graph G_8 admits six round trips, as shown in Listing 3.17.

Listing 3.17: Grounding and solving Program $P_8 \cup P_{15} \cup P_{16}$

```
$ gringo graph.lp costs.lp ham.lp | clasp 0
clasp version 2.0.5
Reading from stdin
Solving...
Answer: 1
cycle(6,3) cycle(5,4) cycle(4,1) cycle(3,5) cycle(2,6) cycle(1,2)
Answer: 2
cycle(6,5) cycle(5,3) cycle(4,1) cycle(3,4) cycle(2,6) cycle(1,2)
Answer: 3
cycle(6,2) cycle(5,6) cycle(4,1) cycle(3,5) cycle(2,4) cycle(1,3)
Answer: 4
cycle(6,3) cycle(5,6) cycle(4,1) cycle(3,4) cycle(2,5) cycle(1,2)
Answer: 5
cycle(6,5) cycle(5,3) cycle(4,2) cycle(3,1) cycle(2,6) cycle(1,4)
Answer: 6
```

---

[2]Such positive recursion makes the resulting ground program non-tight (cf. Section 5.1).

```
cycle(6,3) cycle(5,6) cycle(4,2) cycle(3,1) cycle(2,5) cycle(1,4)
SATISFIABLE

Models      : 6
Time        : 0.001s (Solving: 0.00s 1st Model: 0.00s Unsat: 0.00s)
CPU Time    : 0.000s
```

We have so far ignored edge costs, and stable models of Program P_{16} correspond to round trips only. In order to find a minimum-cost journey among the six round trips of G_8, we add an "optimizing" part to Program P_{16}. This part is expressed by a single minimize statement given in Program P_{17}.

Listing 3.18: Program P_{17} minimizing edge costs for instances of cycle/2 (min.lp)

```
11   #minimize [ cycle(X,Y) = C : cost(X,Y,C) ].
```

Here, edges belonging to the cycle are weighted according to their costs. After grounding, the minimization in Line 11 ranges over 17 instances of cycle/2, one for each (weighted) edge in G_8. For instance, instantiating the weighted (conditional) literal 'cycle(X,Y) = C : cost(X,Y,C)' in view of the fact cost(2,6,4) in Listing 3.15 yields the ground weighted literal 'cycle(2,6) = 4'.

Finally, we explain how the unique minimum-cost round trip (depicted in Figure 3.9) can be computed. The catch is that we are now interested in optimal stable models, rather than arbitrary ones. In order to determine the optimum, we can start by gradually decreasing the costs associated with stable models until we cannot find a strictly better one. In fact, *clasp* successively enumerates better stable models with respect to the provided optimization statements. Any stable model is printed as soon as it has been computed, and the last one is necessarily optimal. If there are multiple optimal stable models, an arbitrary one among them is computed. This proceeding is shown in Listing 3.19.

Listing 3.19: Grounding and solving Program $P_8 \cup P_{15} \cup P_{16} \cup P_{17}$

```
$ gringo graph.lp costs.lp ham.lp min.lp | clasp 0
clasp version 2.0.5
Reading from stdin
Solving...
Answer: 1
cycle(6,3) cycle(5,4) cycle(4,1) cycle(3,5) cycle(2,6) cycle(1,2)
Optimization: 14
Answer: 2
cycle(6,5) cycle(5,3) cycle(4,1) cycle(3,4) cycle(2,6) cycle(1,2)
Optimization: 12
Answer: 3
cycle(6,3) cycle(5,6) cycle(4,1) cycle(3,4) cycle(2,5) cycle(1,2)
Optimization: 11
OPTIMUM FOUND
```

```
Models       : 1
  Enumerated: 3
  Optimum   : yes
Optimization: 11
Time         : 0.001s (Solving: 0.00s 1st Model: 0.00s Unsat: 0.00s)
CPU Time     : 0.000s
```

Given that no answer is obtained after the third one, we know that 11 is the optimum value. However, there might be further stable models sharing the same optimum that have not yet been computed. In order to find them too, we can use the command line option --opt-all=11 to enumerate all stable models having an objective value less or equal to 11. For the graph in Figure 3.8, the optimal stable model is unique. It is illustrated in Figure 3.9.

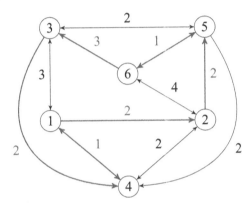

Figure 3.9: A minimum-cost round trip through Graph G_8.

3.4   REFERENCES AND FURTHER READING

The methodology of partitioning a program into a generating, defining, testing, optimizing, and displaying part was coined in Lifschitz (2002). More modeling examples following this paradigm can be found in Eiter et al. (2009), Gebser et al., Leone et al. (2006). Early approaches to modeling in ASP can be found in Baral (2003), Marek and Truszczyński (1999), Niemelä (1999). Modeling aspects from the perspective of Knowledge Representation and Reasoning are discussed in Baral and Gelfond (1994). A best-practice experience is described in Brain et al. (2009).

CHAPTER 4

Grounding

The goal of grounding is to produce a finite and succinct propositional representation of a first-order program. Before addressing the actual grounding process, let us discuss some arising issues.

To begin with, consider Program P_{18}.

$$P_{18} = \left\{ \begin{array}{l} p(a,b) \\ p(b,c) \\ p(X,Z) \leftarrow p(X,Y), p(Y,Z) \end{array} \right\}$$

The signature of P_{18} consists of predicate symbol $p/2$, constant symbols $a/0$, $b/0$, $c/0$, and variable symbols X, Y, Z. Hence, for constructing the ground instantiation $grd(P_{18})$, we have to systematically consider all replacements of the three variables by constants a, b, and c. In the case of the last rule in P_{18} we thus have to deal with twenty-seven ground rules, as witnessed by the exhaustive listing of $grd(P_{18})$ below.

$$grd(P_{18}) = \left\{ \begin{array}{lll} \underline{p(a,b)} & & \\ \underline{p(b,c)} & & \\ p(a,a) \leftarrow p(a,a), p(a,a) & p(b,a) \leftarrow p(b,a), p(a,a) & p(c,a) \leftarrow p(c,a), p(a,a) \\ p(a,a) \leftarrow p(a,b), p(b,a) & p(b,a) \leftarrow p(b,b), p(b,a) & p(c,a) \leftarrow p(c,b), p(b,a) \\ p(a,a) \leftarrow p(a,c), p(c,a) & p(b,a) \leftarrow p(b,c), p(c,a) & p(c,a) \leftarrow p(c,c), p(c,a) \\ p(a,b) \leftarrow p(a,a), p(a,b) & p(b,b) \leftarrow p(b,a), p(a,b) & p(c,b) \leftarrow p(c,a), p(a,b) \\ p(a,b) \leftarrow p(a,b), p(b,b) & p(b,b) \leftarrow p(b,b), p(b,b) & p(c,b) \leftarrow p(c,b), p(b,b) \\ p(a,b) \leftarrow p(a,c), p(c,b) & p(b,b) \leftarrow p(b,c), p(c,b) & p(c,b) \leftarrow p(c,c), p(c,b) \\ p(a,c) \leftarrow p(a,a), p(a,c) & p(b,c) \leftarrow p(b,a), p(a,c) & p(c,c) \leftarrow p(c,a), p(a,c) \\ \underline{p(a,c) \leftarrow p(a,b), p(b,c)} & p(b,c) \leftarrow p(b,b), p(b,c) & p(c,c) \leftarrow p(c,b), p(b,c) \\ p(a,c) \leftarrow p(a,c), p(c,c) & p(b,c) \leftarrow p(b,c), p(c,c) & p(c,c) \leftarrow p(c,c), p(c,c) \end{array} \right\}$$

However, all but one of the twenty-seven rule instances are redundant. For example, rule $p(c,b) \leftarrow p(c,b), p(b,b)$ can never apply, given that there is neither a possibility to derive $p(c,b)$ nor $p(b,b)$. Hence, this rule can be discarded. In fact, $grd(P_{18})$ can equivalently be represented by the two facts and the single rule underlined in $grd(P_{18})$. That is, these three rules yield the same stable models as the twenty-nine listed above.

Next, consider Program P_{19}.

$$P_{19} = \left\{ \begin{array}{l} q(f(a)) \\ p(X) \leftarrow q(X) \end{array} \right\}$$

The signature of P_{19} contains function symbols $a/0$ and $f/1$, leading to the infinite set of terms $\{a, f(a), f(f(a)), \ldots\}$. Consequently, rule $p(X) \leftarrow q(X)$ results in infinitely many ground instantiations in $grd(P_{19})$.

$$grd(P_{19}) \;=\; \left\{ \begin{array}{l} q(f(a)) \\ \underline{p(a) \leftarrow q(a)} \\ \underline{p(f(a)) \leftarrow q(f(a))} \\ p(f(f(a))) \leftarrow q(f(f(a))) \\ p(f(f(f(a)))) \leftarrow q(f(f(f(a)))) \\ \ldots \end{array} \right\}$$

Similar to the above, only one ground instance of $p(X) \leftarrow q(X)$ is essential to $grd(P_{19})$. As a result, the infinite set of rules in $grd(P_{19})$ can be equivalently represented by the two rules underlined above. In contrast to this, we obtain no finite representation of $grd(P_{20})$.

$$P_{20} \;=\; \left\{ \begin{array}{l} q(f(a)) \\ p(X) \leftarrow {\sim}q(X) \end{array} \right\}$$

As above, P_{20} induces the infinite set of terms $\{a, f(a), f(f(a)), \ldots\}$. Unlike the above, however, only one of the infinitely many ground instances of $p(X) \leftarrow {\sim}q(X)$ can be discarded.

$$grd(P_{20}) \;=\; \left\{ \begin{array}{l} q(f(a)) \\ p(a) \leftarrow {\sim}q(a) \\ \underline{p(f(a)) \leftarrow {\sim}q(f(a))} \\ p(f(f(a))) \leftarrow {\sim}q(f(f(a))) \\ \ldots \end{array} \right\}$$

In fact, P_{20} has a single but infinite stable model, which is induced by all rules but $p(f(a)) \leftarrow {\sim}q(f(a))$ in $grd(P_{20})$. For practical reasons, such programs must be rejected.

4.1 BASIC GROUNDING ALGORITHMS

Given a (first-order) logic program P, we are interested in finding a finite subset P' of $grd(P)$ such that the stable models of P' and $grd(P)$ coincide. Although this is often possible, it cannot be determined in general, as we have seen above.

Let us now turn to the computation of finite and succinct representations of ground normal programs (whenever possible). To this end, the concept of safety plays a crucial role: a normal rule is *safe* if each of its variables also occurs in some positive body literal. Accordingly, a normal logic program is safe, if all its rules are safe. Observe that programs P_{18} and P_{19} are safe, while P_{20} is not. Note that a safe normal rule is ground whenever its positive body literals are. In fact, if there are no function symbols of non-zero arity, such a program is guaranteed to have an equivalent finite ground program.

Algorithm 1: NAIVEINSTANTIATION

> **Input**   : A safe (first-order) normal logic program P
> **Output**: A ground normal logic program P'

1 $D := \emptyset$
2 $P' := \emptyset$
3 **repeat**
4 | $D' := D$
5 | **foreach** $r \in P$ **do**
6 | | $B := body(r)^+$
7 | | **foreach** $\theta \in \Theta(B, D)$ **do**
8 | | | $D := D \cup \{head(r)\theta\}$
9 | | | $P' := P' \cup \{r\theta\}$
10 **until** $D = D'$

For a more substantial example, consider the safe program P_{21} (extending P_{18}).

$$P_{21} \;=\; \left\{ \begin{array}{lll} d(a) & p(a,b) & p(X,Z) \leftarrow \underline{p(X,Y)}, p(Y,Z) \\ d(c) & p(b,c) & q(X) \leftarrow \sim r(X), \underline{d(X)} \\ d(d) & p(c,d) & r(X) \leftarrow \sim q(X), \underline{d(X)} \\ q(a) & q(b) & s(X) \leftarrow \sim r(X), \underline{p(X,Y)}, q(Y) \end{array} \right\}$$

Each rule's variables are bound by one of the underlined positive body literals. Given that the set of terms consists of constants only, P_{21} has thus a finite grounding. In contrast, the safe program $\{p(a),\ p(f(X)) \leftarrow p(X)\}$ results in an infinite grounding.

We next provide a simplistic instantiation procedure for safe logic programs. For this, we need the following concepts. A (ground) substitution is a mapping from variables to (ground) terms. Given two sets B and D of atoms, a substitution θ is a *match* of B in D, if $B\theta \subseteq D$. A good match is an inclusion-minimal one because it deals with variables occurring in B only. For instance, both $\{X \mapsto 1\}$ and $\{X \mapsto 2\}$ are good matches of $\{p(X)\}$ in $\{p(1), p(2), p(3)\}$, while $\{X \mapsto 1, Y \mapsto 2\}$ is not a good match. Given a set B of (first-order) atoms and a set D of ground atoms, we define the set $\Theta(B, D)$ of good matches for all elements of B in D as

$$\Theta(B, D) = \{\theta \mid \theta \text{ is a } \subseteq\text{-minimal match of } B \text{ in } D\}.$$

This definition is motivated by the fact that a safe rule is ground once all of its positive body literals are ground.

With the above, we are ready to give our simplistic instantiator in Algorithm 1. The idea is to evaluate a logic program P in a bottom-up fashion by successively building its ground instantiation

P' along with a domain of instantiation D. Let us illustrate Algorithm 1 by building a succinct representation of the ground program of $P_{18} \cup \{p(c, d)\}$.

$$P_{22} = \left\{ \begin{array}{l} p(a, b) \\ p(b, c) \\ p(c, d) \\ p(X, Z) \leftarrow p(X, Y), p(Y, Z) \end{array} \right\}$$

We trace our proceeding in Table 4.1. We initially select rule $p(X, Z) \leftarrow p(X, Y), p(Y, Z)$ upon

$\Theta(B, D)$	D	P'
$\{\emptyset\}$	$p(a, b)$	$p(a, b) \leftarrow$
$\{\emptyset\}$	$p(b, c)$	$p(b, c) \leftarrow$
$\{\emptyset\}$	$p(c, d)$	$p(c, d) \leftarrow$
$\{\{X \mapsto a, Y \mapsto b, Z \mapsto c\},$	$p(a, c)$	$p(a, c) \leftarrow p(a, b), p(b, c)$
$\{X \mapsto b, Y \mapsto c, Z \mapsto d\}\}$	$p(b, d)$	$p(b, d) \leftarrow p(b, c), p(c, d)$
$\{\{X \mapsto a, Y \mapsto c, Z \mapsto d\},$	$p(a, d)$	$p(a, d) \leftarrow p(a, c), p(c, d)$
$\{X \mapsto a, Y \mapsto b, Z \mapsto d\}\}$	$p(a, d)$	$p(a, d) \leftarrow p(a, b), p(b, d)$

Table 4.1: Tracing the instantiation of P_{22}

entering the loop in Line 5. We then get $\Theta(\emptyset, \emptyset) = \emptyset$ because the empty domain offers no match for the positive body literals $p(X, Y)$ and $p(Y, Z)$. Next, consider fact $p(a, b) \leftarrow$. As shown in Table 4.1, we now obtain $\Theta(\emptyset, \emptyset) = \{\emptyset\}$ indicating that the rule can be grounded through the empty substitution. Accordingly, we add atom $p(a, b)$ to the instantiation domain D and rule $p(a, b) \leftarrow$ to the ground program P'. Analogously, facts $p(b, c) \leftarrow$ and $p(c, d) \leftarrow$ are treated upon the next two iterations of the loop in Lines 5–9. This concludes our first traversal of the original program and we re-enter the outer loop in Line 3. This is indicated by a horizontal line in Table 4.1. As before, we start by selecting rule $p(X, Z) \leftarrow p(X, Y), p(Y, Z)$. Unlike the above, however, our instantiation domain has grown to $D = \{p(a, b), p(b, c), p(c, d)\}$. Accordingly, we get for our rule's positive body $\{p(X, Y), p(Y, Z)\}$ the two good matches $\{X \mapsto a, Y \mapsto b, Z \mapsto c\}$ and $\{X \mapsto b, Y \mapsto c, Z \mapsto d\}$. The first match produces the ground rule $p(a, c) \leftarrow p(a, b), p(b, c)$ whose head atom $p(a, c)$ is added to the instantiation domain. Similarly, we produce in the next iteration of the loop in Lines 7–9 the ground rule $p(b, d) \leftarrow p(b, c), p(c, d)$ along with atom $p(b, d)$. The two remaining ground rules are produced in an analogous way. Note that Algorithm 1 produces a ground program consisting of seven rules only. Hence, sixty among the sixty-seven rules in $grd(P_{22})$ turn out to be irrelevant.

Extended rules are instantiated in a similar way. For instance, when instantiating conditional literals like $\ell : \ell_1 : \cdots : \ell_n$, literals $\ell_1, \ldots, \ell_n$ play the role of a rule body. The resulting instances of ℓ are expanded in view of the context of the conditional literal (see Section 2.3.2). Furthermore, the instances of ℓ are added to the instantiation domain D in Algorithm 1 whenever the literal occurs within the head of a rule.

Algorithm 1 is of course an over-simplification of real instantiation procedures. Foremost, it necessitates the re-inspection of the entire program in Line 5. For example by strictly following Algorithm 1, we had to re-inspect all facts upon each iteration. Real implementations, as in the ASP grounders *dlv* or *gringo*, carefully avoid re-grounding rules by using the well-known database technique of semi-naive evaluation. This technique is based on the idea of lazy evaluation. That is, for producing new atoms during an iteration, only rules are considered which have a body atom that was just instantiated in the previous iteration.

Additionally, certain optimizations are conducted during grounding. For example, once a fact like $q(a) \leftarrow$ is encountered when grounding P_{21}, the truth of $q(a)$ is used for partial evaluation. If afterward a rule like $s(b) \leftarrow {\sim}r(b), p(b, a), q(a)$ is about to be generated, it is simplified to $s(b) \leftarrow {\sim}r(b), p(b, a)$. Likewise, rules like $q(a) \leftarrow {\sim}r(a), d(a)$ and $r(a) \leftarrow {\sim}q(a), d(a)$ are skipped right away. With these optimizations the ground program computed in Table 4.1 reduces to seven facts whose heads directly yield the only stable model of P_{22}. Moreover, the use of partial evaluation allows for finitely grounding even larger classes of safe programs. To see this, consider the following program.

$$P_{23} = \left\{ \begin{array}{l} p(a) \\ q(f(f(a))) \\ p(f(X)) \leftarrow p(X), {\sim}q(X) \end{array} \right\}$$

Applying Algorithm 1 together with partial evaluation yields the following finite ground instantiation of P_{23}.

$$P_{23}' = \left\{ \begin{array}{l} p(a) \\ q(f(f(a))) \\ p(f(a)) \leftarrow {\sim}q(a) \\ p(f(f(a))) \leftarrow {\sim}q(f(a)) \end{array} \right\}$$

Although the ground rule $p(f(f(f(a)))) \leftarrow p(f(f(a))), {\sim}q(f(f(a)))$ can be produced, its body is falsified because $q(f(f(a)))$ is true and the rule is skipped. Accordingly, the instantiation domain stays unchanged and no further rules are produced. Similar simplifications taking into account the falsity of atoms can only be done once no more rules with the respective head atom can be produced. This can be accomplished via the graph-theoretic concepts introduced next.

For enabling a more fine-grained static program analysis, we define the *predicate-rule dependency graph* of a logic program P as a directed graph (V, E) where

- V is the set of predicates and rules of P,

- $(p, r) \in E$, if predicate p occurs in the body of rule r, and

- $(r, p) \in E$, if predicate p occurs in the head of rule r.

An edge $(p, r) \in E$ is negative, if predicate p occurs in the negative body of rule r. As an example, consider the predicate-rule dependency graph of Program P_{21} in Figure 4.1. We enclose rules in rectangles and predicates in oval boxes. Negative arcs are indicated by dashed arrows. (Dashed boxes along with their annotated numbers are explained below.)

For capturing the structure within a predicate-rule dependency graph, we take advantage of strongly connected components. In fact, contracting the strongly connected components of a graph into single vertices results in a directed acyclic graph. In view of this, we define a dependency among strongly connected components (U, A) and (W, B) of an encompassing graph (V, E) as follows: (U, A) depends upon (W, B) in (V, E), if $(W \times U) \cap E \neq \emptyset$.

The strongly connected components of the predicate-rule dependency graph of Program P_{21} are shown in Figure 4.1 as dashed boxes. The dependencies among them correspond to the edges

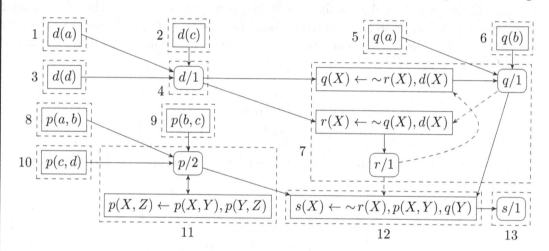

Figure 4.1: Predicate-rule dependency graph of Logic program P_{21} together with a topological order of its strongly connected components.

connecting their contained vertices. We observe that the resulting dependencies are acyclic and thus confirm the above mentioned fact about the contracted graph.

The strongly connected components of a predicate-rule dependency graph are used to partition a logic program. We get 13 such components in the graph given in Figure 4.1. These components can now be separately grounded by following a topological order of their dependencies. That is, we may apply Algorithm 1 in turn to the rules contained in each component while accumulating the instantiation domain. To be more precise, this amounts to making the set D of atoms global and deleting Line 1 in Algorithm 1.

In Figure 4.1 the chosen topological order is indicated via increasing integers next to the components. The instantiation of P_{21} is illustrated in Table 4.2. Once Components 1 to 3 are instantiated, the materialization of predicate $d/1$ is complete. At this point, we know that $d(a), d(c)$,

Table 4.2: Tracing the instantiation of P_{21} following the topological order in Figure 4.1

SCC	$\Theta(B, D)$	D	P'
1	$\{\emptyset\}$	$d(a)$	$d(a) \leftarrow$
2	$\{\emptyset\}$	$d(c)$	$d(c) \leftarrow$
3	$\{\emptyset\}$	$d(d)$	$d(d) \leftarrow$
5	$\{\emptyset\}$	$q(a)$	$q(a) \leftarrow$
6	$\{\emptyset\}$	$q(b)$	$q(b) \leftarrow$
7	$\{\{X \mapsto a\},$ $\{X \mapsto c\},$ $\{X \mapsto d\},$ $\{X \mapsto a\},$ $\{X \mapsto c\},$ $\{X \mapsto d\}\}$	$q(c)$ $q(d)$ $r(c)$ $r(d)$	$q(a) \leftarrow {\sim}r(a), d(a)$ $q(c) \leftarrow {\sim}r(c), d(c)$ $q(d) \leftarrow {\sim}r(d), d(d)$ $r(a) \leftarrow {\sim}q(a), d(a)$ $r(c) \leftarrow {\sim}q(c), d(c)$ $r(d) \leftarrow {\sim}q(d), d(d)$
8	$\{\emptyset\}$	$p(a, b)$	$p(a, b) \leftarrow$
9	$\{\emptyset\}$	$p(b, c)$	$p(b, c) \leftarrow$
10	$\{\emptyset\}$	$p(c, d)$	$p(c, d) \leftarrow$
11	$\{\{X \mapsto a, Y \mapsto b, Z \mapsto c\},$ $\{X \mapsto b, Y \mapsto c, Z \mapsto d\}\}$	$p(a, c)$ $p(b, d)$	$p(a, c) \leftarrow p(a, b), p(b, c)$ $p(b, d) \leftarrow p(b, c), p(c, d)$
	$\{\{X \mapsto a, Y \mapsto c, Z \mapsto d\},$ $\{X \mapsto a, Y \mapsto b, Z \mapsto d\}\}$	$p(a, d)$	$p(a, d) \leftarrow p(a, c), p(c, d)$ $p(a, d) \leftarrow p(a, b), p(b, d)$
12	$\{\{X \mapsto a, Y \mapsto b\},$ $\{X \mapsto a, Y \mapsto c\},$ $\{X \mapsto a, Y \mapsto d\},$ $\{X \mapsto b, Y \mapsto c\},$ $\{X \mapsto b, Y \mapsto d\},$ $\{X \mapsto c, Y \mapsto d\}\}$	$s(a)$ $s(b)$ $s(c)$	$s(a) \leftarrow {\sim}r(a), p(a, b), q(b)$ $s(a) \leftarrow {\sim}r(a), p(a, c), q(c)$ $s(a) \leftarrow {\sim}r(a), p(a, d), q(d)$ $s(b) \leftarrow {\sim}r(b), p(b, c), q(c)$ $s(b) \leftarrow {\sim}r(b), p(b, d), q(d)$ $s(c) \leftarrow {\sim}r(c), p(c, d), q(d)$

and $d(d)$ are true, while $d(b)$ is false. True literals are given in green, false ones in red, all others' truth values are to be determined by a solver. The fixed truth values can now be used for partial evaluation during grounding. For instance, while instantiating Component 7, the previously established truth of $q(a)$ leads to the elimination of rules $q(a) \leftarrow {\sim}r(a), d(a)$ and $r(a) \leftarrow {\sim}q(a), d(a)$. Other rules get simplified. For example, rule $s(b) \leftarrow {\sim}r(b), p(b, c), q(c)$ in Component 12 is reduced by the grounder to $s(b) \leftarrow q(c)$ because ${\sim}r(b)$ and $p(b, c)$ are true. The instantiation of Components 5 and 6 establishes the truth of $q(a)$ and $q(b)$. The truth value of the remaining instances of $q/1$ cannot be determined by the grounder and must be left to the solver. As a consequence, only

---

Algorithm 2: PositiveBodyInstantiation

    Global : Set D of ground atoms
    Input   : A set B of positive body literals
    Output: A set Θ of (ground) substitutions (providing $\Theta(B, D)$)

1   $\Theta := \emptyset$
2   **Procedure** $Instantiation(B', \theta)$
3   **if** $B' = \emptyset$ **then**
4       $\Theta := \Theta \cup \{\theta\}$
5   **else**
6       $b := \text{Select}(B')$
7   **foreach** *good match θ' of $\{b\theta\}$ in* D **do**
8         $Instantiation(B'\backslash\{b\}, \theta \cup \theta')$

9   $Instantiation(B, \emptyset)$

---

rules $q(c) \leftarrow \sim r(c)$ and $q(d) \leftarrow \sim r(d)$ are considered during ASP solving, while facts $q(a) \leftarrow$ and $q(b) \leftarrow$ are merely added to the computed stable models. The remaining components are instantiated in an analogous way. In all, the ground program obtained from P_{21} yields twelve facts and six rules, among which only the latter are subject to ASP solving.

    Finally, let us describe the instantiation of individual rules. In fact, this amounts to solving a constraint satisfaction problem and thus involves search. To this end, we use a simple backtracking approach in Algorithm 2. The algorithm takes the positive body literals B of a rule and outputs $\Theta(B, D)$, namely, all good matches of B in the global domain D. Note that Algorithm 2 builds upon a recursive subroutine (in Lines 2–8) taking a substitution along with a (decreasing) set of positive body literals. We illustrate Algorithm 2 in Table 4.3 by tracing the instantiation of rule $s(X) \leftarrow \sim r(X), p(X, Y), q(Y)$ in Component 12 of P_{21}. The current domain is given by the entries above Component 12 in the column headed D in Table 4.2. Among the positive body literals, we pick $q(Y)$ and match it with $q(a)$. This leads to a recursive call with substitution $\{Y \mapsto a\}$ and remaining body literals $\{p(X, Y)\}$. However, we find no match for $p(X, a)$ in D and must backtrack. Next, we match $q(Y)$ with $q(b)$ and proceed recursively with $\{p(X, Y)\}$ and $\{Y \mapsto b\}$. This time we can match $p(X, b)$ with $p(a, b)$ and obtain $\{X \mapsto a, Y \mapsto b\}$. The remaining substitutions are obtained analogously. As a result, we obtain six good matches for $\{p(X, Y), q(Y)\}$ inducing the six instances of rule $s(X) \leftarrow \sim r(X), p(X, Y), q(Y)$ in Component 12 of P_{21} in Table 4.3.

4.2 TURING MACHINE

Next, let us demonstrate the expressiveness of grounding in the presence of unrestricted function symbols. In fact, the latter makes the language Turing-complete, as we illustrate below by encoding

Table 4.3: Tracing the instantiation of rule $s(X) \leftarrow \sim r(X), p(X, Y), q(Y)$ in Component 12 of P_{21}

$q(Y)$	$p(X, Y)$	S
$\{q(a)$	$\emptyset$	
$q(b),$	$\{p(a, b)\}$	$\{X \mapsto a,\ Y \mapsto b\}$
$q(c),$	$\{p(a, c),$	$\{X \mapsto a,\ Y \mapsto c\}$
	$p(b, c)\}$	$\{X \mapsto b,\ Y \mapsto c\}$
$q(d)\}$	$\{p(a, d),$	$\{X \mapsto a,\ Y \mapsto d\}$
	$p(b, d),$	$\{X \mapsto b,\ Y \mapsto d\}$
	$p(c, d)\}$	$\{X \mapsto c,\ Y \mapsto d\}$

a universal Turing machine in terms of *gringo*'s input language. To this end, we represent a particular instance, namely, a machine solving the 3-state Busy Beaver problem,[1] by the facts in Listing 4.1; its graphical specification is given in Figure 4.2.

Listing 4.1: A 3-state Busy Beaver machine in ASP facts (beaver.lp)

```
1   start(a).
2   blank(0).
3   tape(n,0,n).

5   trans(a,0,1,b,r).
6   trans(a,1,1,c,l).
7   trans(b,0,1,a,l).
8   trans(b,1,1,b,r).
9   trans(c,0,1,b,l).
10  trans(c,1,1,h,r).
```

Fact start(a) and blank(0) specify the starting state a and the blank symbol 0, respectively, of the 3-state Busy Beaver machine. Furthermore, tape(n,0,n) provides the initial tape contents, where 0 indicates a blank at the initial position of the read/write head and the n's represent infinitely many blanks to the left and to the right of the head.[2] Finally, predicate trans/5 captures the transition function of the Busy Beaver machine. A fact of the form trans(S,A,AN,SN,D) describes that, if the machine is in state S and the head is on tape symbol A, it writes AN, changes its state to SN, and moves the head to the left or right according to whether D is l or r.

Listing 4.2 shows an encoding of a universal Turing machine; it defines predicate conf/4 describing the configurations of the machine it runs (for instance, the one specified in Listing 4.1).

---

[1] Our version of the busy beaver machine puts a maximum number of non-blank symbols on the tape.
[2] The fact tape(l(...l(n,s_{-l})...,s_{-1}),s_0,r(s_1,...r(s_r,n)...)) represents the tape s_{-l}...,s_0,...,s_r having the read/write head at symbol s_0.

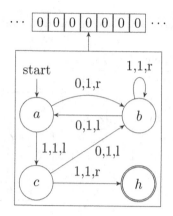

Figure 4.2: A 3-state Busy Beaver machine.

Listing 4.2: An ASP encoding of a universal Turing machine (`turing.lp`)

```
1   conf(S,L,A,R) :- start(S), tape(L,A,R).

3   conf(SN,l(L,AN),AR,R)   :- conf(S,L,A,r(AR,R)), trans(S,A,AN,SN,r).
4   conf(SN,l(L,AN),AR,n)   :- conf(S,L,A,n), blank(AR), trans(S,A,AN,SN,r).
5   conf(SN,L,AL,r(AN,R))   :- conf(S,l(L,AL),A,R), trans(S,A,AN,SN,l).
6   conf(SN,n,AL,r(AN,R))   :- conf(S,n,A,R), blank(AL), trans(S,A,AN,SN,l).
```

The rule in Line 1 determines the starting configuration in terms of a state S, the tape symbol A at the initial position of the read/write head, and the tape contents L and R on its left and right, respectively. The remaining four rules derive successor configurations relative to the transition function (given by facts over `trans/5`). The first two of these rules model movements of the head to the right, thereby distinguishing the cases that the tape contains some (explicit) symbol AR on the right of the head or that its right-hand side is fully blank (n). In the former case, the symbol AN to write is appended to the tape contents on the left of the new head position, represented by means of the functional term l(L,AN), while AR becomes the symbol at the new head position and R the residual contents on its right. Unlike this, the rule dealing with a blank tape on the right takes a blank as the symbol at the new head position and n to represent infinitely many remaining blanks on its right. Similarly, the last two rules specify the symmetric cases obtained for movements to the left. Note that by using function symbols the encoding in Listing 4.2 allows for representing runs of machines without limiting the tape space that can be investigated. Hence, whether *gringo* halts depends on the machine to run. Notably, infinite loops in finite tape space are (implicitly) detected, since repeated configurations do not induce new ground rules.

Invoking *gringo* with files containing the rules in Listing 4.1 and Listing 4.2 yields the following result.

Listing 4.3: Grounding programs `beaver.lp` and `turing.lp`

```
$ gringo --text beaver.lp turing.lp
start(a).
blank(0).
tape(n,0,n).
trans(a,0,1,b,r).
trans(a,1,1,c,l).
trans(b,0,1,a,l).
trans(b,1,1,b,r).
trans(c,0,1,b,l).
trans(c,1,1,h,r).
conf(a,n,0,n).
conf(b,l(n,1),0,n).
conf(a,n,1,r(1,n)).
conf(c,n,0,r(1,r(1,n))).
conf(b,n,0,r(1,r(1,r(1,n)))).
conf(a,n,0,r(1,r(1,r(1,r(1,n))))).
conf(b,l(n,1),1,r(1,r(1,r(1,n)))).
conf(b,l(l(n,1),1),1,r(1,r(1,n))).
conf(b,l(l(l(n,1),1),1),1,r(1,n)).
conf(b,l(l(l(l(n,1),1),1),1),1,n).
conf(b,l(l(l(l(l(n,1),1),1),1),1),0,n).
conf(a,l(l(l(l(n,1),1),1),1),1,r(1,n)).
conf(c,l(l(l(n,1),1),1),1,r(1,r(1,n))).
conf(h,l(l(l(n,1),1),1),1,r(1,n)).
```

In fact, the Turing machine is completely evaluated by *gringo*, which prints all feasible configurations in the same order as a Turing machine would process them. This means that the last line contains the configuration in which the machine reaches the final state. Here, the 3-state Busy Beaver machine terminates after writing six times the symbol 1 to the tape.

The expressive power of Turing-computability should not lead to the idea that the grounder is meant to address computable problems completely by itself. Rather, it provides the most general setting for deterministic computations. In particular, this allows for eliminating many external preprocessing steps involving imperative programming languages.

4.3 META PROGRAMMING

A major consequence of the expressive power of unrestricted function symbols is an easy use of meta modeling techniques in ASP. Following ASP's good practice of uniform encodings, the idea is to re-express logic programs as facts and to combine them with a meta encoding (re-)defining the meaning of the original language constructs. That is, once the original program is reified in terms of facts, it provides data that can be freely treated by the meta encoding. In a way, this allows us to take over the control from the underlying ASP systems and to enforce our own definition of language constructs.

This approach is supported by the grounder *gringo*. To illustrate this, consider the program `easy.lp` in Listing 4.4.

Listing 4.4: A simple program for illustrating meta programming (`easy.lp`)

```
1  { p(1..3) }.
2  :- { p(X) } 2.
3  q(X)  :- p(X),  p(X+1),  X>1.
4  p(X+1)  :- q(X).
```

The human readable output obtained after grounding `easy.lp` (with option `--text`) is given in Listing 4.5; the machine-oriented one is given in Listing 7.5 on Page 121.

Listing 4.5: Grounding Program `easy.lp`

```
0  $ gringo --text easy.lp
1  #count{p(1),p(2),p(3)}.
2  q(2):-p(2),p(3).
3  p(3):-q(2).
4  :-#count{p(3),p(2),p(1)}2.
```

Alternatively, *gringo* offers a reified output format (with option `--reify`). The one obtained after grounding `easy.lp` (and some manual restructuring) is given in Listing 4.6.

Listing 4.6: Reifying Program `easy.lp`

```
0  $ gringo --reify easy.lp

2  rule(pos(sum(0,0,3)),pos(conjunction(0))).
3  rule(pos(atom(q(2))),pos(conjunction(1))).
4  rule(pos(atom(p(3))),pos(conjunction(2))).
5  rule(pos(false),      pos(conjunction(3))).

7  wlist(0,0,pos(atom(p(1))),1).
8  wlist(0,1,pos(atom(p(2))),1).
9  wlist(0,2,pos(atom(p(3))),1).

11 set(1,pos(atom(p(2)))).
12 set(1,pos(atom(p(3)))).

14 set(2,pos(atom(q(2)))).

16 set(3,pos(sum(0,0,2))).

18 scc(0,pos(conjunction(1))).
19 scc(0,pos(atom(q(2)))).
20 scc(0,pos(conjunction(2))).
21 scc(0,pos(atom(p(3)))).
```

Let us explain the reified representation of the ground version of Program `easy.lp` by relating it to its human readable counterpart. The four ground rules in Listing 4.5 are captured by the facts in Lines 2–5 of Listing 4.6. We use the predicate `rule/2` to link rule heads and bodies. By convention, both are positive rule elements, as indicated via the functor pos/1; negative components are furnished with neg/1.

For homogeneity, cardinality constraints are viewed as sum aggregates all of which constituent literals have weight 1 (see Page 21). Thus, the first argument of the fact in Line 2, containing sum(0,0,3), tells us that the head of the captured rule is a sum aggregate. And that it imposes via its first and third argument the trivial bounds 0 and 3 on a list of weighted literals. The second argument of sum/3, here 0, identifies the list of contained weighted literals. The corresponding literals are provided through facts over wlist/4 in Lines 7–9; they all share the first argument 0 as common identifier. While their respective second arguments, 0, 1, and 2, are simply indexes (enabling the representation of duplicates in multisets), the third ones provide the actual literals, p(1), p(2), and p(3), each having the (default) weight 1, as reflected in their fourth arguments.

Again by convention, the body of each rule is a conjunction. For instance, the term conjunction(0) in Line 2 refers to the set (of conjuncts) labeled 0. Given that the rule in Line 1 of Listing 4.5 is a fact, its body is empty, and thus its set of conjuncts is empty, too. This is different with the three other conjunctions in Lines 3–5. For instance, conjunction(1) in Line 3 captures the body consisting of atoms p(2) and p(3). These two conjuncts are represented via the predicate set/2 in Line 11 and 12 of Listing 4.6. Similar to wlist/4, the first argument gives the common identifier, while the second one comprises the actual element. Similarly, Line 14 captures the singleton body of the rule in Line 3 of Listing 4.5.

The fact rule(pos(false),pos(conjunction(3))) in Listing 4.6 accounts for the integrity constraint ':- #count {p(3),p(2),p(1)} 2' in Listing 4.5. This is indicated by the special-purpose constant false. The singleton body contains a positive cardinality constraint with (trivial) lower bound 0 and upper bound 2 over the list labeled 0. This is captured by set(3,pos(sum(0,0,2))) in Line 16. Note that recurrences of lists of weighted literals (and sets) reuse labels introduced before, as done here by referring to 0. This is because *gringo* identifies repetitions of structural entities and reuses labels. In addition to the rules of Listing 4.5, the elements of non-trivial strongly connected components of the corresponding positive dependency graph are obtained in Lines 18–21. To be more precise, the single non-trivial strongly connected component of $G(grd(\texttt{easy.lp}))$ is $(\{q(2),p(3)\}, \{(q(2),p(3)),(p(3),q(2))\})$, where the two edges are established via the rules in Lines 2 and 3 in Listing 4.5. The respective (positive) body elements are represented by the terms pos(atom(q(2))) and pos(conjunction(1)) for the rule in Line 3 as well as pos(atom(p(3))) and pos(conjunction(2)) for the rule in Line 2. The condition that these terms refer to connectors of a common non-trivial strongly connected component, labeled 0, is indicated by the four facts over scc/2 in Lines 18–21. Note that the existence of facts over scc/2 tells us that the program at hand comprises recursion among positive literals. This is referred to as non-tightness (cf. Section 5.1).

Now that we have represented the original program in terms of facts, we can use an encoding for either restoring the original meaning of the program or redefining certain language constructs at will. To see this, consider the "vanilla" meta encoding in Listing 4.7, which simply reestablishes the original meaning of all language constructs.

Listing 4.7: A (restricted) meta encoding (meta.lp)

```
1   litb(B)  :- rule(_,B).
2   litb(E)  :- litb(pos(conjunction(S))), set(S,E).
3   litb(E)  :- eleb(sum(_,S,_)), wlist(S,_,E,_).

5   eleb(P)  :- litb(pos(P)).
6   eleb(N)  :- litb(neg(N)).

8   hold(conjunction(S)) :- eleb(conjunction(S)),
9                 hold(P)       : set(S,pos(P)),
10            not hold(N)       : set(S,neg(N)).

12  hold(sum(L,S,U))       :- eleb(sum(L,S,U)),
13     L #sum [      hold(P) = W : wlist(S,Q,pos(P),W),
14             not hold(N) = W : wlist(S,Q,neg(N),W) ] U.

16  hold(atom(A))          :- rule(pos(atom(A)),   pos(B)), hold(B).

18  L #sum [      hold(P) = W : wlist(S,Q,pos(P),W),
19          not hold(N) = W : wlist(S,Q,neg(N),W) ] U
20                      :- rule(pos(sum(L,S,U)),pos(B)), hold(B).

22                      :- rule(pos(false),     pos(B)), hold(B).

24  #hide.   #show hold(atom(A)).
```

Our meta encoding consists of three parts. The first part in Line 1 to 6 extracts various rule elements. Among them, only those occurring within bodies, identified via eleb/1, are relevant to the generation of stable models specified in the second part in Lines 8 to 22. In fact, stable model generation follows the structure of reified programs. At first, the satisfaction of conjunctions and weight constraints is determined. For example, the rule in Lines 8–10 accounts for conjunctions. For this, it checks whether all associated conjuncts hold. This is done by means of conditional literals, enabling a generic format whose final (ground) form is determined during grounding. Similarly, the rule in Lines 12–14 accounts for weight constraints by inspecting all included weighted literals. For checking the sum of weights of all satisfied literals, the weight, W, of each literals is attached to the corresponding hold/1 atoms. To guarantee that the resulting sum is between the bounds L and U, a built-in #sum aggregate is used. If this condition is satisfied, hold(sum(L,S,U)) is obtained.

Once holding conjunctions and weight constraints are identified, further atoms occurring in rule heads are derived, either singularly or within weight constraints. Line 16 accounts for regular rules, whose head atom is obtained through hold(atom(A)), provided that the corresponding body holds. The same is done in Lines 18–20, yet with rules having a weight constraint in their head. In analogy to Lines 12–14, this is realized via the built-in #sum aggregate. Line 22 deals with integrity constraints represented via the constant false in heads of reified rules.

The last part in Line 24 restricts the output of the meta encoding's stable models to the representations of original input atoms.

Now, we may compute the stable models of our program in Listing 4.4 (easy.lp) by combining the facts in Listing 4.6 with the basic meta encoding in Listing 4.7 (meta.lp). However, before doing so, let us inspect the resulting ground instantiation, given in human readable form in Listing 4.8.[3]

Listing 4.8: Grounding reified Program easy.lp with meta encoding meta.lp

```
0    $ gringo --reify easy.lp | gringo - meta.lp --text

2    wlist(0,0,pos(atom(p(1))),1).
3    wlist(0,1,pos(atom(p(2))),1).
4    wlist(0,2,pos(atom(p(3))),1).
5    rule(pos(sum(0,0,3)),pos(conjunction(0))).
6    set(1,pos(atom(p(2)))).
7    set(1,pos(atom(p(3)))).
8    rule(pos(atom(q(2))),pos(conjunction(1))).
9    set(2,pos(atom(q(2)))).
10   rule(pos(atom(p(3))),pos(conjunction(2))).
11   set(3,pos(sum(0,0,2))).
12   rule(pos(false),pos(conjunction(3))).
13   scc(0,pos(conjunction(1))).
14   scc(0,pos(atom(q(2)))).
15   scc(0,pos(conjunction(2))).
16   scc(0,pos(atom(p(3)))).

18   litb(pos(atom(p(1)))).        eleb(atom(p(1))).
19   litb(pos(atom(p(2)))).        eleb(atom(p(2))).
20   litb(pos(atom(p(3)))).        eleb(atom(p(3))).
21   litb(pos(atom(q(2)))).        eleb(atom(q(2))).
22   litb(pos(conjunction(0))).    eleb(conjunction(0)).
23   litb(pos(conjunction(1))).    eleb(conjunction(1)).
24   litb(pos(conjunction(2))).    eleb(conjunction(2)).
25   litb(pos(conjunction(3))).    eleb(conjunction(3)).
26   litb(pos(sum(0,0,2))).        eleb(sum(0,0,2)).

28   hold(conjunction(0)).
29   hold(conjunction(1)) :- hold(atom(p(3))),hold(atom(p(2))).
30   hold(conjunction(2)) :- hold(atom(q(2))).
31   hold(conjunction(3)) :- hold(sum(0,0,2)).
32   hold(sum(0,0,2)) :-
33          0#sum[hold(atom(p(3)))=1,hold(atom(p(2)))=1,hold(atom(p(1)))=1]2.

35   0#sum[hold(atom(p(3)))=1,hold(atom(p(2)))=1,hold(atom(p(1)))=1]3.
36   hold(atom(q(2))) :- hold(conjunction(1)).
37   hold(atom(p(3))) :- hold(conjunction(2)).
38    :- hold(conjunction(3)).

40   #hide.
```

[3]Following Unix customs, the minus symbol – stands for the output of 'gringo --reify easy.lp'.

```
41   #show hold(atom(p(1))).   #show hold(atom(p(2))).
42   #show hold(atom(q(2))).   #show hold(atom(p(3))).
```

The first part of the obtained ground program consists of the reified program. That is, Lines 2–16 are identical to Listing 4.6 (apart from reordering). The second part results from instantiating the meta encoding in view of the facts in Listing 4.6. Lines 18–26 are a consequence of instantiating Lines 1–6 in Listing 4.7 and thus comprise all rule elements occurring within bodies. The actual encoding of Program easy.lp is given in Lines 28–38. To be more precise, the rules in Lines 28–31 tell us whether the bodies of the rules in Listing 4.5 hold. Analogously, the rule in Line 32/33 indicates whether the cardinality constraint #count {p(3),p(2),p(1)} 2 in the body of Line 4 in Listing 4.5 holds. The remaining rules in Lines 35–38 account for the four ground rules in Listing 4.5 by drawing upon the previously determined status of rule bodies. Finally, Lines 40 to 42 project the resulting stable models on instances of predicate hold/1.

The result of feeding the ground program in Listing 4.8 to the ASP solver *clasp* is shown in Listing 4.9.

Listing 4.9: Solving reified Program easy.lp with meta encoding meta.lp

```
0    $ gringo --reify easy.lp | gringo - meta.lp | clasp
1    clasp version 2.0.5
2    Reading from stdin
3    Solving...
4    Answer: 1
5    hold(atom(q(2))) hold(atom(p(3))) hold(atom(p(2))) hold(atom(p(1)))
6    SATISFIABLE

8    Models      : 1
9    Time        : 0.001s (Solving: 0.00s 1st Model: 0.00s Unsat: 0.00s)
10   CPU Time    : 0.000s
```

In fact, each stable model of the meta encoding applied to the reification of a program corresponds to a stable model of the program being reified. More precisely, a set X of atoms is a stable model of a program iff its meta encoding yields a stable model Y such that $X = \{a \mid \text{hold}(\text{atom}(a)) \in Y\}$. That is, hold(atom($a$)) stands for an atom a.

4.4 REFERENCES AND FURTHER READING

Advanced grounders like *dlv* and *gringo* take advantage of (deductive) database technology, like semi-naive database evaluation. Such techniques can be found in the standard database literature (Abiteboul et al., 1995, Ullman, 1988).

The very first broadly used ASP grounder is the *lparse* system (Syrjänen) written by Tommi Syrjänen; its theoretical foundations are detailed in Syrjänen (2009). Together with the ASP solver *smodels*, it coined the input and intermediate languages used by most ASP systems nowadays, including *gringo* and *clasp*. However, *lparse*'s input language imposed that programs are ω-restricted (Syrjänen, 2001), which boiled down to binding each variable in a rule via a domain predicate. (Intuitively, a predicate is a domain predicate if its domain can be fully evaluated by a grounder; see Gebser et al. (2007d), Syrjänen (2001) for details.) Up to version *3*, *gringo* accepted the slightly extended class of λ-restricted programs (Gebser et al., 2007d). An even more general class is introduced in Lierler and Lifschitz (2009). Beginning with version *3*, *gringo* only requires programs to be safe, similar to the ASP system *dlv* (Leone et al., 2006). The latter as well as its recent upgrade, called *dlv-complex* (Calimeri et al., 2008), integrate grounding and solving capacities. While *dlv*'s solving approach is comparable to that of *smodels*, the underlying grounder is based on similar techniques as used in *gringo* and thus offers the same level of expressiveness. Advanced instantiation techniques can be found in Perri et al. (2007).

The meta programming techniques introduced in Section 4.3 were used in Gebser et al. (2011h) to implemented complex preferences. Similarly, meta interpretation was employed in Eiter et al. (2003) for implementing rule-based preferences, by Eiter and Polleres (2006) for integrating guessing and testing programs, and in Gebser et al. (2008d), Oetsch et al. (2010) for debugging.

CHAPTER 5

Characterizations

This chapter lays the foundations for ASP solving. Thanks to the last chapter, we can from now on restrict ourselves to propositional normal logic programs over sets of ground atoms. For simplicity, we further assume that the underlying set of atoms equals the set of atoms occurring in a program at hand. This makes sense because atoms not appearing in the program are false in any stable model.

In the following sections, we consider alternative characterizations of stable models. Starting from an axiomatic viewpoint, we successively make the underlying inferences more and more precise, ultimately distilling a formal characterization capturing the inferences drawn by the solving algorithms presented in Chapter 6.

5.1 AXIOMATIC CHARACTERIZATION

In order to capture the notion of "negation-as-failure," Clark proposed in (1978) the concept of the *completion* of a logic program. The idea is to capture the semantics of a program with default negation via a translation to classical logic. Although each atom is defined through a set of rules, syntactically, each such rule provides only a sufficient condition for its head atom. The idea of the completion is to turn such implications into a definition by adding the corresponding "necessary" counterpart.

Formally, the *completion* of a logic program P is defined as follows.

$$CF(P) = \left\{ a \leftrightarrow \bigvee_{B \in body_P(a)} BF(B) \mid a \in atom(P) \right\} \tag{5.1}$$

$$\text{where} \quad BF(body(r)) = \bigwedge_{a \in body(r)^+} a \wedge \bigwedge_{a \in body(r)^-} \neg a \tag{5.2}$$

The definition of formula $BF(body(r))$ translates a rule body into a conjunction of literals, while turning default into classical negation. The actual completion $CF(P)$ in (5.1) gathers all bodies implying an atom a within a disjunction, indicating whether one of the rules with head a is applicable. This disjunction is then taken as the definition of atom a. In fact, every stable model of P is a model of $CF(P)$ but not vice versa. An alternative definition is given on Page 85.

For further elaboration, let us partition the completion $CF(P)$ of a program P into two sets of implications, viz. $\overleftarrow{CF}(P)$ and $\overrightarrow{CF}(P)$, as follows.

$$\overleftarrow{CF}(P) = \left\{ a \leftarrow \bigvee_{B \in body_P(a)} BF(B) \mid a \in atom(P) \right\} \tag{5.3}$$

$$\overrightarrow{CF}(P) = \left\{ a \rightarrow \bigvee_{B \in body_P(a)} BF(B) \mid a \in atom(P) \right\} \tag{5.4}$$

Clearly, $CF(P)$ is logically equivalent to $\overleftarrow{CF}(P) \cup \overrightarrow{CF}(P)$. Informally, the implications in $\overleftarrow{CF}(P)$ characterize the classical models of the original program, while the ones in $\overrightarrow{CF}(P)$ are used to complete the program by adding necessary conditions for all atoms.

Models of $\overleftarrow{CF}(P)$ are identical to models of P. The models of $CF(P)$ are called the *supported models* of P. Hence, every stable model of P is a supported model of P. And, by definition, every supported model of P is also a model of P.

For illustration, consider again Program P_7.

$$P_7 = \begin{cases} a \leftarrow \\ b \leftarrow \sim a \\ c \leftarrow a, \sim d \\ d \leftarrow \sim c, \sim e \\ e \leftarrow b, \sim f \\ e \leftarrow e \end{cases}$$

We obtain from P_7 the following sets of implications.

$$\overleftarrow{CF}(P_7) = \begin{cases} a \leftarrow \top \\ b \leftarrow \neg a \\ c \leftarrow a \wedge \neg d \\ d \leftarrow \neg c \wedge \neg e \\ e \leftarrow (b \wedge \neg f) \vee e \\ f \leftarrow \bot \end{cases} \qquad \overrightarrow{CF}(P_7) = \begin{cases} a \rightarrow \top \\ b \rightarrow \neg a \\ c \rightarrow a \wedge \neg d \\ d \rightarrow \neg c \wedge \neg e \\ e \rightarrow (b \wedge \neg f) \vee e \\ f \rightarrow \bot \end{cases}$$

We have seen in Section 2.4 that P_7 has two stable models, namely $\{a, c\}$ and $\{a, d\}$. In contrast, $\overleftarrow{CF}(P_7)$ has twenty-one classical models, among which we find $\{a, c\}$ and $\{a, d\}$ but also $\{a, b, c, d, e, f\}$.

Strengthening $\overleftarrow{CF}(P_7)$ by adding $\overrightarrow{CF}(P_7)$ results in the completion of P_7.

$$CF(P_7) = \begin{cases} a \leftrightarrow \top \\ b \leftrightarrow \neg a \\ c \leftrightarrow a \wedge \neg d \\ d \leftrightarrow \neg c \wedge \neg e \\ e \leftrightarrow (b \wedge \neg f) \vee e \\ f \leftrightarrow \bot \end{cases} \qquad (5.5)$$

In contrast to $\overleftarrow{CF}(P_7)$, the completion formula $CF(P_7)$ has only three classical models, viz. $\{a, c\}$, $\{a, d\}$, and $\{a, c, e\}$. These are the supported models of P_7.

The interesting question is now why the completion fails to characterize the stable models of P_7. For this, let us take a closer look at the only non-stable model of $CF(P_7)$, viz. $\{a, c, e\}$. In fact, the atom e cannot be derived in a non-circular way from P_7. To see this, observe that b is not derivable. Hence, e cannot be derived by rule $e \leftarrow b, \sim f$, which leaves us with $e \leftarrow e$ only, providing a "circular derivation" for e.

Indeed, circular derivations are causing the mismatch between supported and stable models. In fact, all atoms in a stable model can be "derived" from a program in a finite number of steps. Atoms in a cycle (not being "supported from outside the cycle") cannot be "derived" from a program in a finite number of steps. But they do not contradict the completion of a program. Let us make this precise in what follows.

Let X be a stable model of logic program P. Then, for every atom $a \in X$ there is a finite sequence of (positive) rules $\langle r_1, \ldots, r_n \rangle$ such that

1. $head(r_1) = a$,

2. $body(r_i)^+ \subseteq \{head(r_j) \mid i < j \leq n\}$ for $1 \leq i \leq n$,

3. $r_i \in P^X$ for $1 \leq i \leq n$.

That is, each atom of X has a non-circular derivation from P^X. Obviously, there is no finite sequence of rules providing such a derivation for e from $P_7^{\{a,c,e\}}$.

The origin of (potential) circular derivations can be read off the *positive atom dependency graph* of a logic program P given by

$$G(P) = (atom(P), \{(a, b) \mid r \in P, a \in body(r)^+, head(r) = b\}).$$

A logic program P is said to be *tight*, if $G(P)$ is acyclic. Notably, for tight programs, stable and supported models coincide.

Theorem 5.1   *Let P be a tight normal logic program and $X \subseteq atom(P)$.*
    Then, X is a stable model of P iff $X \models CF(P)$.

Looking at the positive atom dependency graph of Program P_7 in Figure 5.1, we observe that P_7 is not tight. This explains the discrepancy among its stable and supported models. Note that eliminating rule $e \leftarrow e$ from P_7 yields a tight program whose stable and supported models

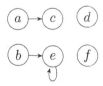

Figure 5.1: Positive atom dependency graph of Logic program P_7.

coincide. For another non-tight example, consider Program P_{24}, whose dependency graph is given in Figure 5.2.

$$P_{24} = \left\{ \begin{array}{llll} a \leftarrow \sim b & c \leftarrow a, b & d \leftarrow a & e \leftarrow \sim a, \sim b \\ b \leftarrow \sim a & c \leftarrow d & d \leftarrow b, c & \end{array} \right\}$$

P_{24} has the two stable models $\{a, c, d\}$ and $\{b\}$. The first stable model is interesting because it

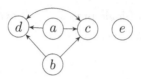

Figure 5.2: Positive atom dependency graph of Logic program P_{24}.

contains c and d, which constitute a cycle in $G(P_{24})$. The reason for this is that d is "externally supported" by $d \leftarrow a$. Such a support is missing for c and d in the second stable model, and thus both atoms are false. On the other hand, $\{b, c, d\}$ is a supported yet not a stable model of P_{24}.

The next interesting question is now whether there is a propositional formula whose models correspond to the stable models of a program. If we consider the completion of a program, $CF(P)$, then the problem boils down to eliminating the circular support of atoms in the supported models of P. The idea is then to add formulas to $CF(P)$ prohibiting circular support of sets of atoms. Note that a circular support between two atoms is possible, whenever there is a path from one atom to another and vice versa in a program's positive atom dependency graph.

This leads us to the concept of a loop.[1] A set $\emptyset \subset L \subseteq atom(P)$ is a *loop* of a logic program P, if it induces a non-trivial strongly connected subgraph of $G(P)$ (the positive atom dependency graph of P). That is, each pair of atoms in L is connected by a path of non-zero length in $(L, E \cap (L \times L))$ where $G(P) = (atom(P), E)$. We denote the set of all loops of P by $loop(P)$. The number of loops in $loop(P)$ may be exponential in $|atom(P)|$. Note that a program P is tight iff $loop(P) = \emptyset$.

In our examples, we get $loop(P_7) = \{\{e\}\}$ and $loop(P_{24}) = \{\{c, d\}\}$, as can be easily verified by looking at Figure 5.1 and 5.2. For an example richer in loops, consider Program P_{25}.

$$P_{25} = \left\{ \begin{array}{llll} a \leftarrow \sim b & c \leftarrow a & d \leftarrow b, c & e \leftarrow b, \sim a \\ b \leftarrow \sim a & c \leftarrow b, d & d \leftarrow e & e \leftarrow c, d \end{array} \right\}$$

The corresponding dependency graph $G(P_{25})$ is given in Figure 5.3. Observe that $\{c, d\}, \{d, e\}$, and $\{c, d, e\}$ are all non-empty sets of atoms whose elements reach one another via paths of non-zero length. Hence, we have that $loop(P_{25}) = \{\{c, d\}, \{d, e\}, \{c, d, e\}\}$.

In fact, we have seen above that the inclusion of a loop in a stable model requires the existence of an external support. This can be made precise as follows. For $L \subseteq atom(P)$, we define the *external*

---

[1]The term loop is used here in a more general way, and not only as an edge connecting a vertex to itself, as common in graph theory.

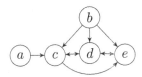

Figure 5.3: Positive atom dependency graphy of Logic program P_{25}.

supports of L in logic program P as

$$ES_P(L) = \{r \in P \mid head(r) \in L, body(r)^+ \cap L = \emptyset\}.$$

For example, we have $ES_{P_{25}}(\{d, e\}) = \{d \leftarrow b, c,\ e \leftarrow b, \sim a\}$. From a technical perspective, however, it is sufficient to consider the respective bodies. Hence, we define the external bodies of L in P as $EB_P(L) = body(ES_P(L))$. We get $EB_{P_{25}}(\{d, e\}) = \{\{b, c\}, \{b, \sim a\}\}$.

With this, we can build formulas to exclude unsupported loops from stable models. The (disjunctive) *loop formula* of L for P is

$$
\begin{aligned}
LF_P(L) &= \left(\bigvee_{a \in L} a\right) \rightarrow \left(\bigvee_{B \in EB_P(L)} BF(B)\right) \\
&\equiv \left(\bigwedge_{B \in EB_P(L)} \neg BF(B)\right) \rightarrow \left(\bigwedge_{a \in L} \neg a\right).
\end{aligned}
$$

The loop formula of L forces all atoms in L to be false whenever L is not externally supported. We define $LF(P) = \{LF_P(L) \mid L \in loop(P)\}$.

The loop $\{e\}$ in Program P_7 induces the loop formula $e \rightarrow b \wedge \neg f$, requiring that e is derived via the rule $e \leftarrow b, \sim f$. In P_{24}, we get $ES_{P_{24}}(\{c, d\}) = \{c \leftarrow a, b,\ d \leftarrow a\}$ yielding the loop formula $c \vee d \rightarrow (a \wedge b) \vee a$, or simplified $c \vee d \rightarrow a$. Similarly, the loop formula $c \vee d \rightarrow a \vee e$ stipulates that the inclusion of loop $\{c, d\}$ into a stable model of P_{25} must be supported by applying $c \leftarrow a$ or $d \leftarrow e$. For loop $\{d, e\}$ of P_{25}, we get the formula $d \vee e \rightarrow (b \wedge c) \vee (b \wedge \neg a)$ or equivalently $(\neg b \vee \neg c) \wedge (\neg b \vee a) \rightarrow \neg d \wedge \neg e$. And finally loop $\{c, d, e\}$ results in the loop formula $c \vee d \vee e \rightarrow a \vee b$.

Adding the loop formulas of a program to its completion provides us with the desired characterization of stable models in terms of classical propositional formulas.

Theorem 5.2   *Let P be a normal logic program and $X \subseteq atom(P)$.*
   Then, X is a stable model of P iff $X \models CF(P) \cup LF(P)$.

In fact, the same result can be obtained with various requirements on loops. A supported model X of a logic program P is a stable model of P if one of the following conditions holds.

- $X \models \{LF_P(U) \mid U \subseteq atom(P)\}$

- $X \models \{LF_P(U) \mid U \subseteq X\}$

- $X \models \{LF_P(L) \mid L \in loop(P)\}$, that is, $X \models LF(P)$

- $X \models \{LF_P(L) \mid L \in loop(P), L \subseteq X\}$

Conversely, if X is not a stable model of P, then there is a loop $L \subseteq X \setminus Cn(P^X)$ such that $X \not\models LF_P(L)$.

To see this, consider the supported model $\{b, c, d\}$ of P_{24}. In fact, we have $Cn(P^{\{b,c,d\}}) = \{b\}$, showing that no atom in the loop $\{c, d\}$ is derivable from $P^{\{b,c,d\}}$. This is also reflected by the fact that the supported model $\{b, c, d\}$ falsifies loop formula $c \vee d \to a$. Unlike this, the loop formula is satisfied by the stable model $\{a, c, d\}$ of P_{24}. And $Cn(P^{\{a,c,d\}}) = \{a, c, d\}$ derives all elements of loop $\{c, d\}$.

To summarize, let us reconsider Program P_7. We have seen on Page 70 that P_7, or equivalently $\overleftarrow{CF}(P_7)$, has twenty-one models. Adding $\overrightarrow{CF}(P_7)$ eliminates eighteen non-supported models. That is, P_7's completion $CF(P_7)$ admits three models, viz. $\{a, c\}$, $\{a, d\}$, and $\{a, c, e\}$. These models constitute the supported models of P_7. The only loop of P_7, viz. $\{e\}$, must be externally supported by rule $e \leftarrow b, \sim f$. We thus have $LF(P_7) = \{e \to b \wedge \neg f\}$. Among the three supported models of P_7, the last one falsifies $LF(P_7)$. Hence, $\{a, c\}$ and $\{a, d\}$ constitute the only models of $CF(P_7) \cup LF(P_7)$ and thus correspond to the two stable models of P_7.

5.2 OPERATIONAL CHARACTERIZATION

We now turn to an operational characterization of supported and stable models in terms of operators on partial interpretations. The interest in such operators lies in the fact that they provide specifications for propagation operations in ASP solvers. To this end, we represent partial interpretations as three-valued interpretations. We define $\langle T_1, F_1 \rangle \sqsubseteq \langle T_2, F_2 \rangle$ if $T_1 \subseteq T_2$ and $F_1 \subseteq F_2$ and $\langle T_1, F_1 \rangle \sqcup \langle T_2, F_2 \rangle = \langle T_1 \cup T_2, F_1 \cup F_2 \rangle$ for three-valued interpretations $\langle T_1, F_1 \rangle$ and $\langle T_2, F_2 \rangle$.

The idea of our first operational characterization is to extend the T_P operator from Section 2.4 to normal logic programs. The idea is to turn a program's completion into an operator, following the implications in (5.3) and the contrapositions of (5.4):

- The head atom of a rule must be *true*, if the rule's body is *true*.

- An atom must be *false*, if the body of each rule having it as head is *false*.

This leads us to the following program-specific operator on partial interpretations.

$$\Phi_P\langle T, F \rangle = \langle \mathbf{T}_P\langle T, F \rangle, \mathbf{F}_P\langle T, F \rangle \rangle \tag{5.6}$$

where

$$\begin{aligned}
\mathbf{T}_P\langle T, F \rangle &= \{head(r) \mid r \in P, body(r)^+ \subseteq T, body(r)^- \subseteq F\} \\
\mathbf{F}_P\langle T, F \rangle &= \{a \in atom(P) \mid body(r)^+ \cap F \neq \emptyset \text{ or } body(r)^- \cap T \neq \emptyset \\
&\qquad \text{for each } r \in P \text{ such that } head(r) = a\}
\end{aligned}$$

We define the iterative variant of Φ_P analogously to T_P as

$$\Phi_P^0 \langle T, F \rangle = \langle T, F \rangle \quad \text{and} \quad \Phi_P^{i+1} \langle T, F \rangle = \Phi_P \Phi_P^i \langle T, F \rangle.$$

As an example, consider Program P_7 from Page 28 along with the iterated application of Φ_{P_7} starting from the "empty" interpretation $\langle \emptyset, \emptyset \rangle$:

$$
\begin{aligned}
\Phi_{P_7}^0 \langle \emptyset, \emptyset \rangle &= & & & \langle \emptyset, \emptyset \rangle \\
\Phi_{P_7}^1 \langle \emptyset, \emptyset \rangle &= & \Phi_{P_7} \langle \emptyset, \emptyset \rangle &= & \langle \{a\}, \{f\} \rangle \\
\Phi_{P_7}^2 \langle \emptyset, \emptyset \rangle &= & \Phi_{P_7} \langle \{a\}, \{f\} \rangle &= & \langle \{a\}, \{b, f\} \rangle \\
\Phi_{P_7}^3 \langle \emptyset, \emptyset \rangle &= & \Phi_{P_7} \langle \{a\}, \{b, f\} \rangle &= & \langle \{a\}, \{b, f\} \rangle \\
\bigsqcup_{i \geq 0} \Phi_{P_7}^i \langle \emptyset, \emptyset \rangle &= & & & \langle \{a\}, \{b, f\} \rangle
\end{aligned}
\tag{5.7}
$$

A partial interpretation $\langle T, F \rangle$ is a Φ_P-fixpoint of P, if $\Phi_P \langle T, F \rangle = \langle T, F \rangle$. The distinguished role of fixpoints is reflected by the following properties.

- The partial interpretation $\bigsqcup_{i \geq 0} \Phi_P^i \langle \emptyset, \emptyset \rangle$ is the $\sqsubseteq$-least Φ_P-fixpoint of P.

  This is because $\Phi_P \langle \emptyset, \emptyset \rangle$ is monotonic, that is, $\Phi_P^i \langle \emptyset, \emptyset \rangle \sqsubseteq \Phi_P^{i+1} \langle \emptyset, \emptyset \rangle$.

- Any other Φ_P-fixpoint extends $\bigsqcup_{i \geq 0} \Phi_P^i \langle \emptyset, \emptyset \rangle$.

- Total Φ_P-fixpoints correspond to supported models of P.

In view of the last statement, Φ_{P_7} has three total fixpoints, whose true atoms correspond to the supported models of P_7, viz. $\{a, c\}$, $\{a, d\}$, and $\{a, c, e\}$. The fact that the truth values of c, d, and e vary explains why the $\sqsubseteq$-smallest fixpoint, $\langle \{a\}, \{b, f\} \rangle$, must leave them undefined.

Nonetheless, Φ_P can be used for approximating stable models of P and so for propagation in ASP solvers. This is because Φ_P is stable-model preserving. That is, whenever $\Phi_P \langle T, F \rangle = \langle T', F' \rangle$, we have $T' \subseteq X$ and $X \cap F' = \emptyset$ for any stable model X of P such that $T \subseteq X$ and $X \cap F = \emptyset$. However, Φ_P is still insufficient because total fixpoints correspond to supported models, not necessarily stable models. Clearly, the problem is the same as with program completion: Φ_P cannot exclude all circular derivations. Nevertheless, for tight programs, Φ_P is sufficient for propagation.

In fact, the operator Φ_P can be strengthened by means of the notion of an unfounded set. Such sets can be viewed as loops lacking any external support and whose atoms thus should become false. A set $U \subseteq atom(P)$ is an *unfounded set* of a logic program P with respect to a partial interpretation $\langle T, F \rangle$ if, for each rule $r \in P$, we have

1. $head(r) \notin U$, or

2. $body(r)^+ \cap F \neq \emptyset$ or $body(r)^- \cap T \neq \emptyset$, or

3. $body(r)^+ \cap U \neq \emptyset$.

Intuitively, $\langle T, F \rangle$ is what we already know about P. Rules satisfying Conditions 1 or 2 are not usable for deriving any atoms in U. Condition 3 is the actual unfounded set condition treating circular derivations: all rules still being usable to derive an atom in U require an(other) atom in U to be true.

For illustration, let us consider the simple program $P = \{a \leftarrow b, b \leftarrow a\}$. Note that P has two supported models, $\emptyset$ and $\{a, b\}$. Hence, the Φ_P-operator cannot assign false to a and b as needed for discarding the non-stable model $\{a, b\}$. However, this can be achieved by identifying $\{a, b\}$ as an unfounded set of P. Let us start by browsing through all candidate unfounded sets. By definition, the empty set is an unfounded set. Furthermore, $\{a\}$ is

- not an unfounded set of P with respect to $\langle \emptyset, \emptyset \rangle$,

- an unfounded set of P with respect to $\langle \emptyset, \{b\} \rangle$, and

- not an unfounded set of P with respect to $\langle \{b\}, \emptyset \rangle$.

While the partial interpretation $\langle \emptyset, \{b\} \rangle$ rules out the only external support of a provided by $a \leftarrow b$, the two others leave this option open. We get the same for $\{b\}$ with roles of a and b reversed. Finally, $\{a, b\}$ is an unfounded set of P with respect to any partial interpretation because both atoms have no support external to $\{a, b\}$.

Similar considerations make us observe that $\{f\}$ is an unfounded set of P_7 with respect to $\langle \emptyset, \emptyset \rangle$; $\{e\}$ and $\{e, f\}$ are unfounded sets of P_7 with respect to $\langle \emptyset, \{b\} \rangle$; $\{b, e\}$ is an unfounded set of P_7 with respect to $\langle \{a\}, \emptyset \rangle$; and $\{b, c, e\}$ is an unfounded set of P_7 with respect to $\langle \{a, d\}, \emptyset \rangle$. And finally we note that $\{c, d\}$ is an unfounded set of P_{24} with respect to $\langle \{b\}, \{a\} \rangle$, but neither with respect to $\langle \emptyset, \emptyset \rangle$ nor $\langle \{a\}, \emptyset \rangle$.

The observation that the union of two unfounded sets is also an unfounded set provides us with a unique set of negative conclusions from a partial interpretation. The *greatest unfounded set* of a logic program P with respect to a partial interpretation $\langle T, F \rangle$, denoted by $\mathbf{U}_P \langle T, F \rangle$, is the union of all unfounded sets of P with respect to $\langle T, F \rangle$. Alternatively, we may define

$$\mathbf{U}_P \langle T, F \rangle = atom(P) \setminus Cn(\{r \in P \mid body(r)^+ \cap F = \emptyset\}^T).$$

Informally, $\mathbf{U}_P \langle T, F \rangle$ gives all atoms that cannot possibly be derived under any circumstances from P in the context of T and F. Observe that $Cn(\{r \in P \mid body(r)^+ \cap F = \emptyset\}^T)$ contains all non-circularly derivable atoms from P with respect to $\langle T, F \rangle$. To see this, consider the following two examples. Taking P_7 and $\langle \{a\}, \emptyset \rangle$ results in

$$Cn(P_7^{\{a\}}) = Cn(\{a \leftarrow, c \leftarrow a, d \leftarrow, e \leftarrow b, e \leftarrow e\}) = \{a, c, d\}$$

from which we get $\mathbf{U}_{P_7} \langle \{a\}, \emptyset \rangle = atom(P_7) \setminus \{a, c, d\} = \{b, e, f\}$.

Analogously, we obtain from P_{24} along with $\langle \{b\}, \{a\} \rangle$,

$$Cn(\{r \in P_{24} \mid body(r)^+ \cap \{a\} = \emptyset\}^{\{b\}}) = Cn(\{b \leftarrow, c \leftarrow d, d \leftarrow c, b\}) = \{b\}$$

and consequently $\mathbf{U}_{P_{24}}\langle\{b\}, \{a\}\rangle = atom(P_{24}) \setminus \{b\} = \{a, c, d, e\}$.

We observe that $\mathbf{U}_{P_7}\langle\{a\}, \emptyset\rangle = \{b, e, f\}$ while $\mathbf{F}_{P_7}\langle\{a\}, \emptyset\rangle = \{b, f\}$, as shown in (5.7) above. In fact, Condition 2 in the definition of an unfounded set corresponds to that of $\mathbf{F}_P$ in operator Φ_P. For a logic program P, this observation leads us to a stronger version of Φ_P by

- keeping the definition of $\mathbf{T}_P$ from Φ_P and

- replacing $\mathbf{F}_P$ in Φ_P by $\mathbf{U}_P$.

In words, an atom must be false, if it belongs to the greatest unfounded set.

As with Φ_P, this results in a program-specific operator on partial interpretations.

$$\Omega_P\langle T, F\rangle = \langle \mathbf{T}_P\langle T, F\rangle, \mathbf{U}_P\langle T, F\rangle\rangle$$

By definition, we have $\Phi_P\langle T, F\rangle \sqsubseteq \Omega_P\langle T, F\rangle$ for any partial interpretation $\langle T, F\rangle$.

As above, we define the iterative variant of Ω_P as

$$\Omega_P^0\langle T, F\rangle = \langle T, F\rangle \quad \text{and} \quad \Omega_P^{i+1}\langle T, F\rangle = \Omega_P \Omega_P^i\langle T, F\rangle.$$

In analogy to (5.7), let us consider the iterated application of Ω_{P_7} starting from the "empty" interpretation $\langle\emptyset, \emptyset\rangle$:

$$
\begin{aligned}
\Omega_{P_7}^0\langle\emptyset, \emptyset\rangle &= & \langle\emptyset, \emptyset\rangle \\
\Omega_{P_7}^1\langle\emptyset, \emptyset\rangle &= \Omega_{P_7}\langle\emptyset, \emptyset\rangle &= \langle\{a\}, \{f\}\rangle \\
\Omega_{P_7}^2\langle\emptyset, \emptyset\rangle &= \Omega_{P_7}\langle\{a\}, \{f\}\rangle &= \langle\{a\}, \{b, e, f\}\rangle \quad (5.8) \\
\Omega_{P_7}^3\langle\emptyset, \emptyset\rangle &= \Omega_{P_7}\langle\{a\}, \{b, e, f\}\rangle &= \langle\{a\}, \{b, e, f\}\rangle \\
\bigsqcup_{i \geq 0} \Omega_{P_7}^i\langle\emptyset, \emptyset\rangle &= \langle\{a\}, \{b, e, f\}\rangle
\end{aligned}
$$

A partial interpretation $\langle T, F\rangle$ is a Ω_P-fixpoint of P, if $\Omega_P\langle T, F\rangle = \langle T, F\rangle$. As with Φ_P above, we have the following properties.

- The partial interpretation $\bigsqcup_{i \geq 0}\Omega_P^i\langle\emptyset, \emptyset\rangle$ is the $\sqsubseteq$-least Ω_P-fixpoint of P.

  This is because $\Omega_P\langle\emptyset, \emptyset\rangle$ is monotonic, that is, $\Omega_P^i\langle\emptyset, \emptyset\rangle \sqsubseteq \Omega_P^{i+1}\langle\emptyset, \emptyset\rangle$.

- Any other Ω_P-fixpoint extends $\bigsqcup_{i \geq 0}\Omega_P^i\langle\emptyset, \emptyset\rangle$.

- Total Ω_P-fixpoints correspond to stable models of P.

In contrast to Φ_{P_7} above, Ω_{P_7} has only two total fixpoints. The true atoms of these fixpoints correspond to the stable models $\{a, c\}$ and $\{a, d\}$ of P_7. Like Φ_P, also Ω_P is stable-model preserving and can therefore be used for approximating stable models of P. Unlike Φ_P, however, operator Ω_P is sufficient for propagation because total fixpoints correspond to stable models. In practice, most ASP solvers apply in addition to Ω_P also backward propagation, originating from program completion (although this is unnecessary from a formal point of view, as we see in the next section).

5.3   PROOF-THEORETIC CHARACTERIZATION

We now introduce a more fine-grained instrument for characterizing inferences in ASP solvers. The idea is to view stable-model computations as derivations in an inference system. To this end, we describe calculi consisting of tableau rules for the construction of stable models of logic programs. A tableau rule captures an elementary inference scheme in an ASP solver. A branch in a tableau corresponds to a successful or unsuccessful computation of a stable model. An entire tableau represents a traversal of the search space.

Inferences in ASP rely on truth values of atoms and applicability of program rules, which can be expressed by assignments over atoms and bodies. For a program P, we thus fix the domain of assignments A to $dom(A) = atom(P) \cup body(P)$. Such a hybrid approach may result in exponentially smaller tableaux (and thus search space traversals) than either a purely atom- or body-based approach, as shown at the end of this section.

A tableau for a logic program P and an initial assignment A is a binary tree with the rules of P and the entries of A at its root.[2] Further nodes in the tree are restricted to entries of the form $\boldsymbol{T}v$ or $\boldsymbol{F}v$ for $v \in dom(A)$. They are generated by applying tableau rules in the standard way: given a tableau rule and a branch in a tableau such that the prerequisites of the rule hold in the branch, the tableau can be extended by appending entries to the end of the branch as specified by the rule. Note that every branch corresponds to a pair (P, A). We draw on this relationship for identifying branches below.

The tableau rules for normal programs P are shown in Figure 5.4. For convenience, they make use of two conjugation functions, $\boldsymbol{t}$ and $\boldsymbol{f}$. For an entry ℓ, define:

$$\boldsymbol{t}\ell = \begin{cases} \boldsymbol{T}\ell & \text{if } \ell \in dom(A) \\ \boldsymbol{F}v & \text{if } \ell = {\sim}v \text{ for } v \in dom(A) \end{cases}$$

$$\boldsymbol{f}\ell = \begin{cases} \boldsymbol{F}\ell & \text{if } \ell \in dom(A) \\ \boldsymbol{T}v & \text{if } \ell = {\sim}v \text{ for } v \in dom(A) \end{cases}$$

In view of this, the *FTB* rule in *(a)* expresses that truth of a rule body can be deduced if the body's literals hold in a branch. Conversely, if the body is already assigned to false and all but one literal hold, the remaining literal must necessarily be false; this contrapositive argument is formalized by the *BFB* rule in *(b)*. Likewise, the tableau rules *FTA* and *FFB* in *(c)* and *(e)* capture straightforward conditions under which an atom must be assigned to true and a body to false, respectively. Their contrapositives are given by *BFA* and *BTB* in *(d)* and *(f)*. The remaining tableau rules in *(g)*–*(k)* are subject to provisos. For an application of *FFA* in *(g)*, deducing an unsupported atom a to be false, (§) stipulates that $B_1, \ldots, B_m$ comprise all bodies of rules with head a. Its contrapositive, the *BTA* rule in *(h)*, is also guided by (§). The outer structure of *WFN* [Ξ] and *WFJ* [Ξ] in *(i)* and *(j)*, aiming at unfounded sets, is similar to *FFA* and *BTA*, yet their proviso (†[Ξ]) requires a concerned atom a to belong to some set $U \in \Xi$ such that $B_1, \ldots, B_m$ comprise all external bodies

---

[2]We refrain from marking rules in P by $\boldsymbol{T}$ as they are not subject to an assignment through A.

$$\frac{a \leftarrow \ell_1, \ldots, \ell_n}{\boldsymbol{t}\ell_1, \ldots, \boldsymbol{t}\ell_n} \over \boldsymbol{T}\{\ell_1, \ldots, \ell_n\}}$$

(a) Forward True Body (FTB)

$$\frac{\boldsymbol{F}\{\ell_1, \ldots, \ell_{i-1}, \ell_i, \ell_{i+1}, \ldots, \ell_n\}}{\boldsymbol{t}\ell_1, \ldots, \boldsymbol{t}\ell_{i-1}, \boldsymbol{t}\ell_{i+1}, \ldots, \boldsymbol{t}\ell_n} \over \boldsymbol{f}\ell_i}$$

(b) Backward False Body (BFB)

$$\frac{a \leftarrow \ell_1, \ldots, \ell_n}{\boldsymbol{T}\{\ell_1, \ldots, \ell_n\}} \over \boldsymbol{T}a}$$

(c) Forward True Atom (FTA)

$$\frac{a \leftarrow \ell_1, \ldots, \ell_n}{\boldsymbol{F}a} \over \boldsymbol{F}\{\ell_1, \ldots, \ell_n\}}$$

(d) Backward False Atom (BFA)

$$\frac{a \leftarrow \ell_1, \ldots, \ell_i, \ldots, \ell_n}{\boldsymbol{f}\ell_i} \over \boldsymbol{F}\{\ell_1, \ldots, \ell_i, \ldots, \ell_n\}}$$

(e) Forward False Body (FFB)

$$\frac{\boldsymbol{T}\{\ell_1, \ldots, \ell_i, \ldots, \ell_n\}}{\boldsymbol{t}\ell_i}$$

(f) Backward True Body (BTB)

$$\frac{\boldsymbol{F}B_1, \ldots, \boldsymbol{F}B_m}{\boldsymbol{F}a} \ (\S)$$

(g) Forward False Atom (FFA)

$$\frac{\boldsymbol{T}a \quad \boldsymbol{F}B_1, \ldots, \boldsymbol{F}B_{i-1}, \boldsymbol{F}B_{i+1}, \ldots, \boldsymbol{F}B_m}{\boldsymbol{T}B_i} \ (\S)$$

(h) Backward True Atom (BTA)

$$\frac{\boldsymbol{F}B_1, \ldots, \boldsymbol{F}B_m}{\boldsymbol{F}a} \ (\dagger[\Xi])$$

(i) Well-Founded Negation (WFN[\Xi])

$$\frac{\boldsymbol{T}a \quad \boldsymbol{F}B_1, \ldots, \boldsymbol{F}B_{i-1}, \boldsymbol{F}B_{i+1}, \ldots, \boldsymbol{F}B_m}{\boldsymbol{T}B_i} \ (\dagger[\Xi])$$

(j) Well-Founded Justification (WFJ[\Xi])

$$\frac{}{\boldsymbol{T}v \ \mid \ \boldsymbol{F}v} \ (\sharp[\Gamma])$$

(k) Cut (Cut[\Gamma])

$$
\begin{array}{lll}
(\S) & : & a \in atom(P), body_P(a) \subseteq \{B_1, \ldots, B_m\} \subseteq body(P) \\
(\dagger[\Xi]) & : & a \in U, U \in \Xi, \ EB_P(U) \subseteq \{B_1, \ldots, B_m\} \subseteq body(P) \\
(\sharp[\Gamma]) & : & v \in \Gamma
\end{array}
$$

Figure 5.4: Tableau rules for normal programs.

of U in P. Two alternatives of interest for Ξ are $\Xi = 2^{atom(P)}$ and $\Xi = loop(P)$. Finally, $(\sharp[\Gamma])$ guides applications of the $Cut[\Gamma]$ rule in *(k)* by restricting (choice) variables v to members of Γ. For a normal program P, we below consider different sets $\Gamma \subseteq atom(P) \cup body(P)$.[3] Note that a Cut application adds entries Tv and Fv as the left and the right child to the end of a branch, thus reflecting non-determinism in assigning v. With every other tableau rule, its consequent is appended to a branch, that is, applications are deterministic.

For illustration consider Program P_1. An example (complete) tableau for P_1 is given in Figure 5.5. The applications of tableau rules are indicated by their names, e.g., *(FTB)* and $(Cut[atom(P_1)])$, respectively. We observe that both branches in Figure 5.5 comprise P_1 (at the root) along with total assignments over $atom(P_1) \cup body(P_1)$: the left branch represents stable model $\{a, c\}$, as indicated by Ta and Tc, while the right one gives stable model $\{a, d\}$.

$$a \leftarrow$$
$$c \leftarrow \sim b, \sim d$$
$$d \leftarrow a, \sim c$$

		$T\emptyset$		*(FTB)*
		Ta		*(FTA)*
		Fb		*(FFA)*
Tc		Fc		$(Cut[atom(P_1)])$
$T\{\sim b, \sim d\}$	*(BTA)*	$F\{\sim b, \sim d\}$	*(BFA)*	
Fd	*(BTB)*	Td	*(BFB)*	
$F\{a, \sim c\}$	*(FFB)*	$T\{a, \sim c\}$	*(FTB)*	

Figure 5.5: Complete tableau for P_1 and the empty assignment.

Let us now turn to the characterization of deterministic consequences. For some $v \in dom(A)$, we say that Tv or Fv can be deduced by a set $\mathcal{T}$ of tableau rules in a branch (P, A) if the entry can be generated by applying some rule in $\mathcal{T}$ other than Cut. Accordingly, we let $D_{\mathcal{T}}(P, A)$ denote the set of entries deducible by $\mathcal{T}$ in (P, A). Moreover, $D_{\mathcal{T}}^*(P, A)$ represents the set of entries in a smallest branch that extends (P, A) and is closed under $\mathcal{T}$, that is, $D_{\mathcal{T}}(P, D_{\mathcal{T}}^*(P, A)) \subseteq D_{\mathcal{T}}^*(P, A)$. For an example, let us consider Program P_7 along with an empty assignment. For $\mathcal{T}$ we simply take all tableau rules in Figure 5.4 except for Cut. The stepwise formation of deterministic consequences is illustrated in Figure 5.6. In formal terms, we thus get

$$D_{\mathcal{T}}^*(P_7, \emptyset) = \{Ff, T\emptyset, Ta, F\{\sim a\}, Fb, F\{b, \sim f\}, Fe, F\{e\}\}. \tag{5.9}$$

Looking at assigned atoms only, we observe that the derived consequences correspond to those obtained by the iterated application of operator Ω_{P_7} in (5.8).

---

[3]The Cut rule may, in principle, introduce more general entries; this would however necessitate additional decomposition rules, leading to extended tableau calculi.

$$a \leftarrow$$
$$b \leftarrow \sim a$$
$$c \leftarrow a, \sim d$$
$$d \leftarrow \sim c, \sim e$$
$$e \leftarrow b, \sim f$$
$$e \leftarrow e$$

$$
\begin{array}{ll}
\boldsymbol{F}f & (FFA) \\
\boldsymbol{T}\emptyset & (FTB) \\
\boldsymbol{T}a & (FTA) \\
\boldsymbol{F}\{\sim a\} & (FFB) \\
\boldsymbol{F}b & (FFA) \\
\boldsymbol{F}\{b, \sim f\} & (FFB) \\
\boldsymbol{F}e & (WFN\,[2^{\,atom(P_7)}]) \\
\boldsymbol{F}\{e\} & (FFB)
\end{array}
$$

Figure 5.6: Tableau branch for P_7 and the empty assignment.

Note that all deterministic tableau rules in Figure 5.4 are stable-model preserving; this also applies to the *Cut* rule when considering both resulting branches. Different tableau calculi, viz. particular rule sets, yield characteristic correspondences. For some examples, consider the following.[4]

$$
\begin{array}{rcl}
\Phi_P \langle T, F \rangle & \sim & D^*_{\{FTB,FTA,FFB,FFA\}}(P, A) \\
\Omega_P \langle T, F \rangle & \sim & D^*_{\{FTB,FTA,FFB,WFN\,[2^{atom(P)}]\}}(P, A) \\
Unit\,Propagation(CF(P) \cup C_A) & \sim & D^*_{\{(a)-(h)\}}(P, A) \\
expand'_P(L, U) & \sim & D^*_{\{(a)-(h),WFN\,[2^{atom(P)}]\}}(P, A)
\end{array}
$$

Apart from the respective representation of assignments and interpretations,[5] we see that the difference between operator Φ_P and Ω_P manifests itself in the choice between tableau rule *FFA* and *WFN*. A more fine-grained analysis shows that $\mathbf{T}_P$, $\mathbf{F}_P$, and $\mathbf{U}_P$ are captured by $\{FTA, FTB\}$, $\{FFA, FFB\}$, and $\{WFN\,[2^{atom(P)}], FFB\}$, respectively. Note that *FTB* and *FFB* serve merely as intermediate propagators for body assignments, while the specific atom-wise assignments are done by *FTA*, *FFA*, and $WFN\,[2^{atom(P)}]$, respectively. Moreover, unit propagation from the completed program — abbreviated by $Unit\,Propagation(CF(P) \cup C_A)$; see Page 83 for more details[6] — can be seen as an extension of operator Φ_P by backward propagation, viz. tableau rules *BTB*, *BTA*,

---

[4]These correspondences are not exact in the sense that the given tableau calculi deal with both atoms and bodies and thus allow for more inferences than obtainable in the purely atom-based approaches on the left.

[5]That is, we leave the exact correspondence of A to $\langle T, F \rangle$, C_A, and (L, U) implicit.

[6]We use C_A as an informal placeholder capturing the assignment A.

BFB, and *BFA*. Analogously, propagation as accomplished by the ASP solver *smodels*, referred to as[7] *expand$'_P$* (see also Page 32), amounts to an enhancement of Ω_P by backward propagation.

Finally, let us see how tableau calculi allow us to characterize stable model computations. A branch (P, A) is contradictory if A is contradictory, and non-contradictory otherwise. Moreover, (P, A) is complete (with respect to a tableau calculus $\mathcal{T}$) if it is contradictory or if A is total and $D_{\mathcal{T}}(P, A) \subseteq A$. A tableau is complete if all its branches are complete. A complete tableau for a logic program and the empty assignment such that all branches are contradictory is called a refutation for the program (meaning that the program has no stable model). As an example, let us consider the tableau calculus comprising the tableau rules $\{(a) - (h), WFN\ [2^{atom(P)}]\}$ along with *Cut*[*atom(P)*]: given a normal logic program P, we have the following characterizations of stable models. [8]

1. P has a stable model X iff every complete tableau for P and $\emptyset$ has a unique non-contradictory branch (P, A) such that $A^T \cap atom(P) = X$.

2. P has no stable model iff every complete tableau for P and $\emptyset$ is a refutation.

Note that instead of *Cut*[*atom(P)*], also *Cut*[*body(P)*] and *Cut*[*atom(P)* $\cup$ *body(P)*] are sufficient to complete tableaux for P and $\emptyset$. However, different proof complexities are obtained with respect to such *Cut* variants. In fact, the proof system obtained with *Cut*[*atom(P)* $\cup$ *body(P)*] is exponentially stronger than the ones with either *Cut*[*atom(P)*] or *Cut*[*body(P)*]. The practical consequence of this is that ASP solvers permitting both atoms and bodies as choice variables may traverse exponentially smaller search spaces.

5.4 NOGOOD-BASED CHARACTERIZATION

We finally develop a uniform constraint-based framework capturing the whole spectrum of inferences in ASP. This paves the way for harnessing advanced Boolean constraint technology for implementing ASP solving. For representing such constraints, we take advantage of the concept of a *nogood*. This allows us to view inferences in ASP as unit propagation.

In our setting, a *nogood* is a set $\{\sigma_1, \ldots, \sigma_m\}$ of entries, expressing that any assignment containing $\sigma_1, \ldots, \sigma_m$ is inadmissible. Accordingly, a total assignment A is a *solution* for a set Δ of nogoods if $\delta \not\subseteq A$ for all $\delta \in \Delta$.

For instance, given the domain $\{a, b\}$, the total (unordered) assignment $\{Ta, Fb\}$ is a solution for the nogoods $\{Ta, Tb\}$ and $\{Fa, Fb\}$. Likewise, $\{Fa, Tb\}$ is another solution. Importantly, nogoods provide us with reasons explaining why entries must belong to a solution, and look-back techniques can be used to analyze and recombine inherent reasons for conflicts (see Chapter 6 for details).

In fact, deterministic tableau rules like the ones on Page 79 inherently induce nogoods, given that such rules express the fact that their prerequisites necessarily imply their consequent. That is, a

---

[7]We use *expand$'_P$* in order to stress the extension to *expand$_P$* in Section 2.4.

[8]As this calculus admits a (unique) non-contradictory complete branch (P, A) in some tableau iff (P, A) belongs to every complete tableau for P and $\emptyset$, the statements remain valid when replacing "every" by "some."

tableau rule with prerequisites $\sigma_1, \ldots, \sigma_n$ and consequent σ expresses the fact that $\{\sigma_1, \ldots, \sigma_n, \overline{\sigma}\}$ is a nogood. Investigating all instances of deterministic tableau rules for a logic program P thus allows for extracting a set Δ of nogoods such that any solution A for Δ corresponds to a non-contradictory complete branch (P, A) in a tableau (and vice versa).

Once all inferences are captured in terms of nogoods, they can be drawn in a uniform way by means of unit propagation: given a nogood δ and an assignment A, an entry $\sigma \notin A$ is *unit-resulting* for δ with respect to A, if $\delta \setminus A = \{\overline{\sigma}\}$. That is, if all but one entry of a nogood are contained in an assignment, the complement of the remaining entry must hold in any solution extending the current assignment. For a set Δ of nogoods and an assignment A, *unit propagation* is the iterated process of extending A with unit-resulting entries until no further entry is unit-resulting for any nogood in Δ.[9]

To illustrate this, let us take up the above example. Given the partial (unordered) assignment $\{Ta\}$ along with the nogood $\{Ta, Tb\}$, we observe that all entries in the nogood but Tb are already contained in the assignment. Hence, any solution extending the assignment $\{Ta\}$ must exclude Tb since otherwise it would contain the entire nogood $\{Ta, Tb\}$. In turn, all eligible extensions of assignment $\{Ta\}$ must contain Fb. This is expressed by the fact that Fb is unit-resulting for nogood $\{Ta, Tb\}$ with respect to the assignment $\{Ta\}$. Unit propagation then extends assignment $\{Ta\}$ by Fb, leading to the augmented assignment $\{Ta, Fb\}$. On the other hand, no entry is unit-resulting for nogood $\{Fa, Fb\}$ with respect to assignment $\{Ta\}$.

The specification of Boolean constraints given below follows the axiomatic characterization of stable models in Section 5.1. This definition distinguishes between the completion, $CF(P)$, and the loop formulas, $LF(P)$, of a normal logic program P. In fact, models of $CF(P)$ match non-contradictory complete branches in tableaux containing the deterministic tableau rules *(a)–(h)* in Figure 5.4. Furthermore, we have seen in Section 5.3 that augmenting these rules with $WFN[2^{atom(P)}]$ (or equivalently $WFN[loop(P)]$) characterizes models of $CF(P) \cup LF(P)$, that is, the stable models of P. The major difference between $CF(P)$ and $LF(P)$ is that the former captures local conditions applying to individual atoms and rule bodies, while $LF(P)$ aims at more global conditions related to unfounded sets.

In the following, we specify nogoods such that their solutions correspond to stable models of a given program. To begin with, the set of *completion nogoods*, Δ_P, of a normal program P is defined as follows.

$$\Delta_P \;=\; \bigcup_{B \in body(P), B = \{\ell_1, \ldots, \ell_n\}} \left\{ \begin{array}{l} \{FB, t\ell_1, \ldots, t\ell_n\}, \\ \{TB, f\ell_1\}, \ldots, \{TB, f\ell_n\} \end{array} \right\} \qquad (5.10)$$

$$\cup \bigcup_{a \in atom(P), body_P(a) = \{B_1, \ldots, B_k\}} \left\{ \begin{array}{l} \{Ta, FB_1, \ldots, FB_k\}, \\ \{Fa, TB_1\}, \ldots, \{Fa, TB_k\} \end{array} \right\} \qquad (5.11)$$

The completion nogoods Δ_P can be derived by decomposing the completion $CF(P)$ into a set of clauses. To be more precise, the first set of body-oriented nogoods in (5.10) is obtained from

---

[9]In view of the proof-theoretic considerations at the end of the previous section, we keep considering entries over both atoms and bodies of an underlying program.

$BF(body(r))$ in (5.2), while the second set of atom-oriented ones in (5.11) can be derived from $CF(P)$ in (5.1). The major difference is that bodies are taken above as first-class objects while they are dealt with implicitly in Section 5.1. See below for illustration.

The set Δ_P can also be obtained by converting the tableau rules *(a)–(h)* in Figure 5.4 into nogoods. To be more precise, the nogood $\{FB, t\ell_1, \ldots, t\ell_n\}$ expresses the fact that a body B must not be false if all of its entries hold; the same exclusion is achieved by tableau rule *FTB* (or *BFB*, provided that $B \neq \emptyset$). The nogoods $\{TB, f\ell_1\}, \ldots, \{TB, f\ell_n\}$, representing that B cannot hold if one of its entries is false, comply with tableau rule *FFB* or, alternatively, *BTB*. If B is empty, that is, if $n = 0$, there are no nogoods of this kind, and the corresponding tableau rules are likewise inapplicable. Turning to an atom a, the nogood $\{Ta, FB_1, \ldots, FB_k\}$ denies solutions containing a while all its supporting bodies are false. Tableau rule *FFA* (or *BTA*, provided that $body_P(a) \neq \emptyset$) expresses the same. Finally, the nogoods $\{Fa, TB_1\}, \ldots, \{Fa, TB_k\}$ view program rules as implications, complying with tableau rule *FTA* or, alternatively, *BFA*; if a has no supporting rule, that is, if $k = 0$, there are no nogoods of this kind, and the corresponding tableau rules are likewise inapplicable.

As an example, consider the following normal logic program.

$$P_{26} = \begin{cases} a &\leftarrow& \sim b \\ b &\leftarrow& \sim a \\ c &\leftarrow& a \\ c &\leftarrow& d \\ d &\leftarrow& c, \sim a \\ e &\leftarrow& c \\ e &\leftarrow& d \end{cases}$$

We consider entries and thus nogoods over $atom(P_{26}) \cup body(P_{26})$. The tableau rules *(a)–(h)* in Figure 5.4 correspond to the nogoods in $\Delta_{P_{26}}$ as shown in Table 5.1.

Table 5.1: Set $\Delta_{P_{26}}$ of nogoods and associated tableau rules	
Tableau Rules	Nogoods in $\Delta_{P_{26}}$
FTB, BFB	$\{F\{\sim b\}, Fb\}$, $\{F\{\sim a\}, Fa\}$, $\{F\{a\}, Ta\}$, $\{F\{d\}, Td\}$, $\{F\{c\}, Tc\}$, $\{F\{c, \sim a\}, Tc, Fa\}$
FFB, BTB	$\{T\{\sim b\}, Tb\}$, $\{T\{\sim a\}, Ta\}$, $\{T\{a\}, Fa\}$, $\{T\{d\}, Fd\}$, $\{T\{c\}, Fc\}$, $\{T\{c, \sim a\}, Fc\}$, $\{T\{c, \sim a\}, Ta\}$
FTA, BFA	$\{Fa, T\{\sim b\}\}$, $\{Fb, T\{\sim a\}\}$, $\{Fc, T\{a\}\}$, $\{Fc, T\{d\}\}$, $\{Fd, T\{c, \sim a\}\}$, $\{Fe, T\{c\}\}$, $\{Fe, T\{d\}\}$
FFA, BTA	$\{Ta, F\{\sim b\}\}$, $\{Tb, F\{\sim a\}\}$, $\{Tc, F\{a\}, F\{d\}\}$, $\{Td, F\{c, \sim a\}\}$, $\{Te, F\{c\}, F\{d\}\}$

Since each body occurring in P_{26} is non-empty and each atom has a supporting rule, every nogood in $\Delta_{P_{26}}$ reflects exactly one forward- and one backward-oriented tableau rule. The nogoods in $\Delta_{P_{26}}$ can be derived systematically by considering potential applications of the tableau rules *(a)–(h)* in Figure 5.4 for each target variable $v \in atom(P_{26}) \cup body(P_{26})$. As a consequence, a nogood like $\{\boldsymbol{F}\{c, \sim a\}, \boldsymbol{T}c, \boldsymbol{F}a\}$ captures three distinct inferences, one for each of its (complementary) entries. They correspond to the following (instances of) tableau rules.

$$
\begin{array}{ccc}
\begin{array}{c} d \leftarrow c, \sim a \\ \boldsymbol{T}c \\ \boldsymbol{F}a \\ \hline \boldsymbol{T}\{c, \sim a\} \end{array}
&
\begin{array}{c} \boldsymbol{F}\{c, \sim a\} \\ \boldsymbol{F}a \\ \hline \boldsymbol{F}c \end{array}
&
\begin{array}{c} \boldsymbol{F}\{c, \sim a\} \\ \boldsymbol{T}c \\ \hline \boldsymbol{T}a \end{array}
\\[2ex]
(a)\ FTB & (b)\ BFB & (b)\ BFB
\end{array}
$$

In terms of unit propagation, these rules express that the entry in the consequent is unit-resulting whenever the entries in the prerequisite belong to the current assignment.

Similarly, nogood $\{\boldsymbol{T}c, \boldsymbol{F}\{a\}, \boldsymbol{F}\{d\}\}$ corresponds to the following three tableau rules.

$$
\begin{array}{cccc}
\begin{array}{c} \boldsymbol{F}\{a\} \\ \boldsymbol{F}\{d\} \\ \hline \boldsymbol{F}c \end{array}\ (\S)
&
\begin{array}{c} \boldsymbol{T}c \\ \boldsymbol{F}\{a\} \\ \hline \boldsymbol{T}\{d\} \end{array}\ (\S)
&
\begin{array}{c} \boldsymbol{T}c \\ \boldsymbol{F}\{d\} \\ \hline \boldsymbol{T}\{a\} \end{array}\ (\S)
&
(\S) : body_{P_{26}}(c) = \{\{a\}, \{d\}\}
\\[2ex]
(g)\ FFA & (h)\ BTA & (h)\ BTA &
\end{array}
$$

Note that all three inferences are sanctioned by the proviso expressed in (§).

For realizing the direct relationship between the completion nogoods and the actual completion of a program, it is sufficient to re-express the latter by using auxiliary atoms turning bodies into first-class objects. Alternatively, the completion of a logic program P can be defined with the help of a set of auxiliary variables for bodies, viz. $\{v_B \mid B \in body(P)\}$.

$$
\begin{aligned}
CF^x(P) \quad = \quad & \{v_B \leftrightarrow (\bigwedge_{a \in body(r)^+} a \wedge \bigwedge_{a \in body(r)^-} \neg a) \mid r \in P\} \\
& \cup \{a \leftrightarrow (\bigvee_{B \in body_P(a)} v_B) \mid a \in atom(P)\}
\end{aligned}
$$

Let us illustrate this by showing the correspondence between $\Delta_{P_{26}}$ and the completion $CF^x(P_{26})$ of P_{26}. The (alternative) completion $CF^x(P_{26})$ is as follows.

$$
CF^x(P_{26}) \quad = \quad \left\{
\begin{array}{llll}
a & \leftrightarrow & v_{\{\sim b\}} & \quad v_{\{\sim b\}} \leftrightarrow \neg b \\
b & \leftrightarrow & v_{\{\sim a\}} & \quad v_{\{\sim a\}} \leftrightarrow \neg a \\
c & \leftrightarrow & v_{\{a\}} \vee v_{\{d\}} & \quad v_{\{a\}} \leftrightarrow a \\
& & & \quad v_{\{d\}} \leftrightarrow d \\
d & \leftrightarrow & v_{\{c, \sim a\}} & \quad v_{\{c, \sim a\}} \leftrightarrow c \wedge \neg a \\
e & \leftrightarrow & v_{\{c\}} \vee v_{\{d\}} & \quad v_{\{c\}} \leftrightarrow c
\end{array}
\right\}
$$

The models of $CF^x(P_{26})$ and corresponding solutions for $\Delta_{P_{26}}$ are shown in Table 5.2. The equiv-

Model of $CF(P_{26})$	Model of $CF^x(P_{26})$	Solution for $\Delta_{P_{26}}$
$\{a, c, e\}$	$\{a, c, e\} \cup$ $\{v_{\{\sim b\}}, v_{\{a\}}, v_{\{c\}}\}$	$\{Ta, Fb, Tc, Fd, Te\} \cup$ $\{T\{\sim b\}, F\{\sim a\}, T\{a\}, F\{d\},$ $T\{c\}, F\{c, \sim a\}\}$
$\{b\}$	$\{b\} \cup$ $\{v_{\{\sim a\}}\}$	$\{Fa, Tb, Fc, Fd, Fe\} \cup$ $\{F\{\sim b\}, T\{\sim a\}, F\{a\}, F\{d\},$ $F\{c\}, F\{c, \sim a\}\}$
$\{b, c, d, e\}$	$\{b, c, d, e\} \cup$ $\{v_{\{\sim a\}}, v_{\{d\}}, v_{\{c\}}, v_{\{c, \sim a\}}\}$	$\{Fa, Tb, Tc, Td, Te\} \cup$ $\{F\{\sim b\}, T\{\sim a\}, F\{a\}, T\{d\},$ $T\{c\}, T\{c, \sim a\}\}$

Table 5.2: Models of $CF(P_{26})$, $CF^x(P_{26})$, and corresponding solutions for $\Delta_{P_{26}}$

alences in $CF^x(P_{26})$ can be divided in the atom-oriented ones on the left and the body-oriented ones on the right. This reflects the division of tableau rules into atom- and body-oriented rules in Figure 5.4. For an atom-oriented example, consider the formula $c \leftrightarrow v_{\{a\}} \vee v_{\{d\}}$. It is equivalent to the clauses

$$c \vee \neg v_{\{a\}}, \quad c \vee \neg v_{\{d\}}, \quad \text{and} \quad \neg c \vee v_{\{a\}} \vee v_{\{d\}} .$$

In turn, negating each clause and substituting each auxiliary variable v_B by B yields the nogoods $\{Fc, T\{a\}\}$, $\{Fc, T\{d\}\}$, and $\{Tc, F\{a\}, F\{d\}\}$. Similarly, the body-oriented formula $v_{\{c, \sim a\}} \leftrightarrow c \wedge \neg a$ is equivalent to the clauses

$$\neg v_{\{c, \sim a\}} \vee c, \quad \neg v_{\{c, \sim a\}} \vee \neg a, \quad \text{and} \quad v_{\{c, \sim a\}} \vee \neg c \vee a .$$

Proceeding in analogy to the above, we obtain the nogoods $\{T\{c, \sim a\}, Fc\}$, $\{T\{c, \sim a\}, Ta\}$, and $\{F\{c, \sim a\}, Tc, Fa\}$.

Accordingly, we have the following counterpart to Theorem 5.1 characterizing stable models of tight programs in terms of nogoods.

Theorem 5.3   *Let P be a tight normal logic program and $X \subseteq atom(P)$.*
   Then, X is a stable model of P iff $X = A^T \cap atom(P)$ for a (unique) solution A for Δ_P.

In order to extend the characterization of stable models by solutions for nogoods to non-tight programs, P, we additionally need to reflect tableau rule $WFN[2^{atom(P)}]$ or $WFN[loop(P)]$ from Figure 5.4. To this end, we define the set of *loop nogoods* of P, Λ_P, as follows.

$$\Lambda_P = \bigcup_{U \subseteq atom(P), EB_P(U) = \{B_1, \dots, B_k\}} \{\{Ta, FB_1, \dots, FB_k\} \mid a \in U\} \qquad (5.12)$$

The nogoods in Λ_P express that an atom a must not be true if it belongs to an unfounded set U. The same exclusion is achieved by tableau rule $WFN[2^{atom(P)}]$ (or $WFJ[2^{atom(P)}]$), provided that

$EB_P(U) \neq \emptyset$). Note that the definition in (5.12) is rather general in selecting subsets $U \subseteq atom(P)$. An alternative accommodating $WFN[loop(P)]$ is to select U among $loop(P)$.

In view of the correspondence between (deterministic) tableau rules in Figure 5.4 and $\Delta_P \cup \Lambda_P$, we derive the following counterpart to Theorem 5.2 in terms of nogoods.

Theorem 5.4   *Let P be a normal logic program and $X \subseteq atom(P)$.*
Then, X is a stable model of P iff $X = A^T \cap atom(P)$ for a (unique) solution A for $\Delta_P \cup \Lambda_P$.

Let us reconsider P_{26} along with the solutions for $\Delta_{P_{26}}$ in Table 5.2. In fact, P_{26} is not tight because $loop(P_{26}) = \{\{c, d\}\}$ due to rules $c \leftarrow d$ and $d \leftarrow c, \sim a$. In view of $EB_{P_{26}}(\{c, d\}) = \{\{a\}\}$, we obtain for $U = \{c, d\}$ the loop nogoods $\{Tc, F\{a\}\}$ and $\{Td, F\{a\}\}$. In analogy to Table 5.1, these nogoods capture all inferences obtainable from $WFN[loop(P_{26})]$ and $WFJ[loop(P_{26})]$. Now, checking the three solutions for $\Delta_{P_{26}}$ in Table 5.2, we observe that the third one contains both loop nogoods, while neither is included in the first two solutions. In other words, the first two solutions of $\Delta_{P_{26}}$ are also solutions for $\Delta_{P_{26}} \cup \Lambda_{P_{26}}$, while the third one violates $\Lambda_{P_{26}}$. According to Theorem 5.4, the corresponding sets of true atoms, $\{a, c, e\}$ and $\{b\}$, respectively, are the stable models of P_{26}.

With the full set of nogoods at hand, we can now also provide the counterpart of the tableau branch in Figure 5.6 in terms of unit propagation (see also (5.9)). The result is shown in Figure 5.7. The left column gives the unit-resulting entry, σ, for the nogood, δ, in the mid-

σ	δ	$\Delta_{P_7} \cup \Lambda_{P_7}$
Ff	$\{Tf\}$	Δ_{P_7}
$T\emptyset$	$\{F\emptyset\}$	Δ_{P_7}
Ta	$\{Fa, T\emptyset\}$	Δ_{P_7}
$F\{\sim a\}$	$\{T\{\sim a\}, Ta\}$	Δ_{P_7}
Fb	$\{Tb, F\{\sim a\}\}$	Δ_{P_7}
$F\{b, \sim f\}$	$\{T\{b, \sim f\}, Fb\}$	Δ_{P_7}
Fe	$\{Te, F\{b, \sim f\}\}$	Λ_{P_7}
$F\{e\}$	$\{T\{e\}, Fe\}$	Δ_{P_7}

Figure 5.7: Unit propagation on $\Delta_{P_7} \cup \Lambda_{P_7}$.

dle column with respect to the previously obtained entries. At the same time, the left column develops the growing assignment downward. The rightmost column indicates whether the propagating nogood is a completion or a loop nogood. In total, we thus constructed the assignment $(Ff, T\emptyset, Ta, F\{\sim a\}, Fb, F\{b, \sim f\}, Fe, F\{e\})$ by unit propagation on $\Delta_{P_7} \cup \Lambda_{P_7}$. Recall that the assigned atoms correspond to the deterministic consequences obtained by the iterated application of operator Ω_{P_7} in (5.8) on Page 77.

By Theorem 5.4, the nogoods in $\Delta_P \cup \Lambda_P$ describe a set of constraints that need to be checked for identifying stable models. However, while the size of Δ_P is linear in the size of P, the size of Λ_P is in general exponential. This is however no defect in the construction of Λ_P but an implication of (widely accepted assumptions in) complexity theory. Hence, most ASP solvers work on logic programs as succinct representations of loop nogoods (or formulas, respectively) and check them efficiently by determining unfounded sets relative to assignments. See Chapter 6 for details.

Although we do not detail this here, let us illustrate the generality of the constraint-based approach by sketching how it captures weight constraints. As with rule bodies, the constraint-based characterization is twofold. First, the assignment's domain is extended so that weight constraints can be assigned truth values. Second, the set of inferences induced by weight constraints is described in terms of a set of nogoods.

As an example, consider the weight constraint on Page 22:

$$\omega = 10 \, \{course(db) = 6, course(ai) = 6, course(project) = 8, course(xml) = 3\} \, 20$$

In this case, the Boolean variable ω is introduced and the above weight constraint is captured by the set of nogoods in Figure 5.8. The nogoods in Δ_ω containing $F\omega$ express (minimal) combinations

$$\Delta_\omega = \left\{ \begin{array}{l} \{F\omega, T course(db), T course(ai), F course(project)\}, \\ \{F\omega, T course(db), T course(ai), F course(xml)\}, \\ \{F\omega, T course(db), T course(project), F course(ai)\}, \\ \{F\omega, T course(db), T course(project), F course(xml)\}, \\ \{F\omega, T course(ai), T course(project), F course(db)\}, \\ \{F\omega, T course(ai), T course(project), F course(xml)\}, \\ \{F\omega, T course(project), T course(xml), F course(db)\}, \\ \{F\omega, T course(project), T course(xml), F course(ai)\}, \\ \{T\omega, F course(db), F course(ai), F course(xml)\}, \\ \{T\omega, F course(db), F course(project)\}, \\ \{T\omega, F course(ai), F course(project)\}, \\ \{T\omega, T course(db), T course(ai), T course(project), T course(xml)\} \end{array} \right\}$$

Figure 5.8: Nogoods for an example weight constraint.

of entries over atoms in ω such that the lower and the upper bound of ω are definitely satisfied, so that ω must not be assigned F. Similarly, the first three nogoods with $T\omega$ provide combinations of entries under which the lower bound of ω cannot be established, and the last nogood expresses that the upper of ω is violated when all atoms in ω are assigned T. Taken as a whole, the set Δ_ω of nogoods requires that the truth value assigned to ω matches the valuation of ω with respect to its constituents. Conditions that rely on the valuation of ω, e.g., concerning rule bodies that include ω, can then be formulated transparently by referring to $T\omega$ or $F\omega$, respectively.

5.5   REFERENCES AND FURTHER READING

Clark introduced with the concept of completion in (1978) the first declarative account of negation-as-failure, or in a broader sense, of closed world reasoning.

Non-circular derivations are a distinguishing feature of ASP's forerunner, Default Logic (Reiter, 1980) (in contrast to Autoepistemic Logic (Moore, 1985)). This feature was first described by Schwind (1990) and referred to as groundedness. In ASP, the conceptually different property of tightness was identified by Fages in (1994). As well, Theorem 5.1 was shown by Fages (1994), providing a first step toward the characterization of stable models in terms of propositional formulas in classical logic.

Lin and Zhao present in (2004) the fundamental characterization of stable models in terms of propositional formulas (cf. Theorem 5.2; see also Section 1.2 on limitations of translations between ASP and SAT). Lin (1991) characterizes stable models of logic programs in terms of Circumscription (McCarthy, 1980, 1986); see also Pearce et al. (2009). The concepts of loops and loop formulas were identified in Lin and Zhao (2004) and further elaborated in Lee (2005). Unfounded sets are linked to loops, as they provide a syntactic characterization of "interesting" unfounded sets. Gebser and Schaub (2005), Gebser et al. (2011i) identify elementary loops that can be used to confine unfounded set checking to necessary parts.

Operator Φ_P was conceived by Fitting in (1985), and is thus often referred to as the *Fitting operator*. With it, we may define the *Fitting semantics* of a logic program P as the partial interpretation $\bigsqcup_{i \geq 0} \Phi_P^i \langle \emptyset, \emptyset \rangle$. The Fitting semantics of P is not conflicting and generally not total.

The operator Ω_P was invented by Van Gelder et al. (1991) along with the central concept of an unfounded set. See Alviano et al. (2011), Leone et al. (1997) for more literature on unfounded sets. The *well-founded semantics* of a logic program P is defined as the partial interpretation $\bigsqcup_{i \geq 0} \Omega_P^i \langle \emptyset, \emptyset \rangle$. As the Fitting semantics, the well-founded semantics of P is not conflicting and generally not total. However, the well-founded semantics yields more consequences than Fitting's, that is, $\bigsqcup_{i \geq 0} \Phi_P^i \langle \emptyset, \emptyset \rangle \sqsubseteq \bigsqcup_{i \geq 0} \Omega_P^i \langle \emptyset, \emptyset \rangle$ for any program P.

As pointed out by Van Gelder (1993), we can define an anti-monotonic operator from the basic ingredients of the original definition of stable models, namely, $\Psi_P(X) = Cn(P^X)$. With it, the stable models of a program P can be characterized through fixpoints of operator Ψ_P. Given that Ψ_P is anti-monotonic, its squared variant Ψ_P^2 is monotonic. As discussed in Lifschitz (1996), all stable models of P include all (well-founded) atoms belonging to the smallest fixpoint of Ψ_P^2, while they exclude all (unfounded) atoms belonging to the complement of the greatest fixpoint of Ψ_P^2. For more details on such alternating fixpoints, the interested reader is referred to Lifschitz (1996), Van Gelder (1993).

A comprehensive introduction to tableau systems is provided by the Tableau Handbook (D'Agostino et al., 1999). The tableau-based characterization of ASP solving given in Section 5.3 was introduced in Gebser and Schaub (2006a) and extended in Gebser and Schaub (2007), Järvisalo and Oikarinen (2008); see also Gebser and Schaub (2012). As pointed out in Hähnle (2001), DPLL is very similar to the propositional version of KE tableau calculus (D'Agostino and Mondadori, 1994). Further related work includes tableau calculi for other forms of Nonmonotonic Reasoning (Dix et al., 2001, Olivetti, 1999, Pearce et al., 2000). An alternative transition-based framework for characterizing DPLL- and CDCL-based search procedures for SAT solving was introduced by Nieuwenhuis et al. (2006). This framework was employed by Lierler (2011) to characterize search procedures for ASP solving. General investigations into propositional proof complexity (Cook and Reckhow, 1979), in particular, that of (UN)SAT, can be found in (Beame and Pitassi, 1998). Recent results on CDCL (Beame et al., 2004, Pipatsrisawat and Darwiche, 2011) indicate its relation to general resolution.

Nogoods constitute a traditional concept in Constraint Processing (CP) for characterizing constraints in a complementary way (see Dechter (2003), Rossi et al. (2006)). Of course, clauses can be viewed as particular syntactic representations of nogoods, just as other representations, like gates, inequalities, rules, etc. Therefore, we employ nogoods for abstracting from the respective syntactic entities in order to obtain a canonical means of characterization.

The notion of a unit-resulting entry is closely related to that of a unit clause in SAT solving (cf. Biere et al. (2009)). Along the lines of SAT, we call the iterated process of extending an assignment by unit-resulting entries unit propagation. The nogood-based characterization of stable models given in Section 5.4 was introduced in Gebser et al. (2007a); see also Gebser et al. (2012b). This characterization was extended to disjunctive logic programs in Drescher et al. (2008) and to weight constraints in Gebser et al. (2009a). A more general study is given in Gebser and Schaub (2006b).

Further characterizations of stable models can be found in Lifschitz (2008).

CHAPTER 6

Solving

The nogood-based characterization developed in Section 5.4 provides us with a uniform constraint-based framework for different kinds of inferences in ASP. In particular, it allows us to implement inferences in ASP as unit propagation on nogoods and thus to take advantage of advanced Boolean constraint technology. To begin with, let us consider the two most popular algorithmic schemes in Boolean Constraint Solving.

6.1 BOOLEAN CONSTRAINT SOLVING

The *solve* algorithm developed in Section 2.4 (on Page 29) follows the scheme of the well-known *Davis-Putnam-Logemann-Loveland* (DPLL) algorithm. This algorithm was developed about five decades ago and constitutes the traditional approach to SAT solving. The outline of DPLL is given in Figure 6.1. The idea is to combine deterministic (unit) propagation with systematic backtracking,

loop

    propagate   // compute deterministic consequences

    if no conflict **then**

        if all variables assigned **then return** variable assignment

        else *decide*   // non-deterministically assign some literal

    else

        if top-level conflict **then return** unsatisfiable

        else

            backtrack   // undo assignments made after last decision

           flip   // assign complement of last decision literal

Figure 6.1: Basic decision algorithm for backtracking based Boolean constraint solving (DPLL).

in case of a conflict flipping the last non-deterministically assigned literal. Although the presentation of DPLL differs from the one of *solve* in Section 2.4 (because Figure 6.1 is aligned with Figure 6.2), it follows the same pattern. At first, all deterministic consequences are computed. Then, three cases are distinguished. First, if all variables have been consistently assigned, the computation terminates and outputs the stable model represented by the variable assignment. Second, if only some variables have been consistently assigned (and the remaining ones are still unassigned), we choose a variable and assign it a truth value. This assignment is a non-deterministic consequence. Third, if some variables have been assigned inconsistently, we backtrack and reassign the previously chosen variable the complementary truth value. A top-level conflict designates the special case in which a conflict is obtained from deterministic consequences only (and only made explicit here to stay compatible with Figure 6.2).

The search pattern of modern SAT solvers is referred to as *Conflict-Driven Clause Learning* (CDCL). It is outlined in Figure 6.2. Like DPLL, it starts with the computation of deterministic

loop

    propagate   // compute deterministic consequences

    if no conflict **then**

        if all variables assigned **then return** variable assignment

        else *decide*   // non-deterministically assign some literal

    else

        if top-level conflict **then return** unsatisfiable

        else

            analyze   // analyze conflict and add a conflict constraint

           backjump   // undo assignments until conflict constraint is unit

Figure 6.2: Basic decision algorithm for conflict-driven Boolean constraint learning (CDCL).

consequences, followed by the above described case analysis. However, the basic idea of CDCL follows the principle of "learning from mistakes." Rather than merely undoing the last choice, CDCL starts with an analysis of the encountered conflict that aims at determining its origin. Once this is accomplished, CDCL enriches the original problem specification by learning from the encountered conflict and then returns to the source of this conflict. The major technical difference between CDCL and DPLL lies in the look-back techniques utilized to recover from conflicts: CDCL applies an *analyze* step that strengthens the input by adding a conflict constraint; it also performs a *backjump* to a point where the conflict constraint is unit (yields some deterministic consequence by propagation). That is, CDCL replaces (i) systematic backtracking by backjumping, and (ii) flips of former decision literals by inferences through conflict constraints. We illustrate this in the sequel.

Next, let us instantiate the algorithmic scheme of CDCL with the nogood-based concepts of Section 5.4 in order to accommodate ASP solving. In fact, once a logic program has been converted into nogoods, the above CDCL scheme can be adapted in a straightforward way. However, as motivated in Section 5.5, we present our conflict-driven learning algorithm for deciding stable

---

Algorithm 3: CDNL-ASP

Input   : A normal program P.
Output : A stable model of P or "no stable model."

1 $A := \emptyset$ // assignment over $atom(P) \cup body(P)$
2 $\nabla := \emptyset$ // set of recorded nogoods
3 $dl := 0$ // decision level

4 **loop**
5 $\quad (A, \nabla) := \text{NogoodPropagation}(P, \nabla, A)$ // deterministic consequences
6 $\quad$ **if** $\varepsilon \subseteq A$ **for some** $\varepsilon \in \Delta_P \cup \nabla$ **then** // conflict
7 $\quad\quad$ **if** $\max(\{dlevel(\sigma) \mid \sigma \in \varepsilon\} \cup \{0\}) = 0$ **then return** no stable model
8 $\quad\quad (\delta, dl) := \text{ConflictAnalysis}(\varepsilon, P, \nabla, A)$
9 $\quad\quad \nabla := \nabla \cup \{\delta\}$ // (temporarily) record conflict nogood
10 $\quad\quad A := A \setminus \{\sigma \in A \mid dl < dlevel(\sigma)\}$ // backjumping
11 $\quad$ **else if** $A^T \cup A^F = atom(P) \cup body(P)$ **then**   // stable model
12 $\quad\quad$ **return** $A^T \cap atom(P)$
13 $\quad$ **else**
14 $\quad\quad \sigma_d := \text{Select}(P, \nabla, A)$ // decision
15 $\quad\quad dl := dl + 1$
16 $\quad\quad dlevel(\sigma_d) := dl$
17 $\quad\quad A := A \circ \sigma_d$

---

models existence in terms of nogoods and, in the following, call it *Conflict-Driven Nogood Learning for ASP (CDNL-ASP)*. The result is given as Algorithm 3 and explained in what follows.

6.2   SETTING THE STAGE

While the order of entries in a Boolean assignment does not affect its semantics, it is crucial for algorithms analyzing the entries' interdependencies. Hence, we now consider ordered assignments like $A = (\sigma_1, \ldots, \sigma_n)$. We associate for $1 \leq i \leq n$ with each entry σ_i in A a non-negative integer $dlevel(\sigma_i)$, called the *decision level* of σ_i. Furthermore, we let $A[\sigma_i] = (\sigma_1, \ldots, \sigma_{i-1})$ denote the prefix of A relative to σ_i, and define $A[\sigma] = A$ if $\sigma \notin A$. Finally, we require that in any ordered assignment distinct entries assign distinct variables. This implies that ordered assignments are non-contradictory.

For an ordered assignment $A = (\sigma_1, \ldots, \sigma_n)$ and an entry σ with an associated decision level, $dlevel(\sigma)$, we let $A \circ \sigma$ stand for the result of appending σ to A, that is, $(\sigma_1, \ldots, \sigma_n, \sigma)$ provided that $\max(\{dlevel(\sigma_1), \ldots, dlevel(\sigma_n)\} \cup \{0\}) \leq dlevel(\sigma)$. That is, $A \circ \sigma$ contains σ as the last entry with a decision level greater than or equal to $dlevel(\sigma_n)$. For instance, appending $\boldsymbol{F}d$ with $dlevel(\boldsymbol{F}d) = 2$ to $A = (\boldsymbol{T}a, \boldsymbol{F}b, \boldsymbol{T}c)$ with $dlevel(\boldsymbol{T}a) = 0$, $dlevel(\boldsymbol{F}b) = 1$, and $dlevel(\boldsymbol{T}c) = 1$ yields $A \circ \boldsymbol{F}d = (\boldsymbol{T}a, \boldsymbol{F}b, \boldsymbol{T}c, \boldsymbol{F}d)$ along with all previous decision levels. Hence, decision levels are increasing along the sequence of entries in an ordered assignment.

The concepts of unit-resulting entries and unit propagation carry over from unordered to ordered assignments in the straightforward way. For instance, $\boldsymbol{F}d$ is unit-resulting for the nogood $\{\boldsymbol{F}b, \boldsymbol{T}d\}$ with respect to assignment $(\boldsymbol{T}a, \boldsymbol{F}b, \boldsymbol{T}c)$, but it is neither with respect to $(\boldsymbol{T}a, \boldsymbol{F}b, \boldsymbol{T}c, \boldsymbol{T}d)$ nor $(\boldsymbol{T}a, \boldsymbol{F}b, \boldsymbol{T}c, \boldsymbol{F}d)$. For an ordered assignment A and an entry σ, we call a nogood δ an antecedent of σ with respect to A if σ is unit-resulting for δ with respect to $A[\sigma]$. Hence, in the previous example, $\{\boldsymbol{F}b, \boldsymbol{T}d\}$ is an antecedent of $\boldsymbol{F}d$ with respect to assignment $(\boldsymbol{T}a, \boldsymbol{F}b, \boldsymbol{T}c)$. Extending this to sets Δ of nogoods, we say that σ is implied by Δ with respect to A, if Δ contains some antecedent of σ with respect to A. The notion of implication identifies entries that must necessarily be added to an assignment because the addition of their complement would violate some nogood. Implied entries are crucial for the meaningful application of conflict analysis, described in Section 6.6.

When given a normal program P, throughout this chapter, we assume that all variables occurring in an associated set of nogoods belong to $atom(P) \cup body(P)$; in particular, this applies to dynamic nogoods, which are denoted by ∇ below.

6.3 CONFLICT-DRIVEN NOGOOD LEARNING

We now present the basic decision algorithm for conflict-driven Boolean nogood learning. At this abstract level, it applies to any set of nogoods, no matter whether they stem from a logic program, a set of clauses, or any other Boolean domain.

Our main procedure for finding a stable model (or deciding that there is none) is shown in Algorithm 3. Given a normal program P, the algorithm starts from an empty assignment A and an empty set ∇ of recorded nogoods over $atom(P) \cup body(P)$. The latter set is used to accumulate conflict and loop nogoods. The recorded nogoods in ∇ and the completion nogoods in Δ_P are exploited for (unit) propagation and conflict analysis. Moreover, we use the decision level dl for counting the number of *decision entries* in A. The initial value of dl is 0. Decision entries are non-deterministically chosen (in Line 14 in Algorithm 3), while entries derived (deterministically) by propagation (in Line 5) are implied by $\Delta_P \cup \nabla$ with respect to A.

For computing a stable model of P, the main loop in Lines 4–16 of Algorithm 3 follow the standard proceeding of CDCL. First, NogoodPropagation (detailed in Section 6.4) deterministically extends A in Line 5, and possibly also records loop nogoods from Λ_P in ∇. Afterward, one of the following cases applies:

Conflict. If propagation led to the violation of some nogood $\varepsilon \in \Delta_P \cup \nabla$, as checked in Line 6, there are two possibilities.

- Either the conflict occurred independently of any previous decision, meaning that the input program P has no stable model, or

- ConflictAnalysis (cf. Section 6.6) is performed in Line 8 to determine a conflict nogood δ, recorded in ∇ in Line 9, along with a decision level dl to jump back to.

Note that we assume δ to be *asserting*, that is, some entry must be unit-resulting for δ after backjumping in Line 10. This condition is guaranteed by ConflictAnalysis and makes sure that after backjumping, Algorithm 3 traverses the search space differently than before (without explicitly flipping any decision entry).

Solution. If propagation led to a total assignment A (not violating any nogood in $\Delta_P \cup \nabla$), as checked in Line 11, the atoms true in A belong to a stable model of P and are returned in Line 12.

Decision. If neither of the previous cases applies, A is partial, and a decision entry σ_d is selected according to some heuristic in Line 14. We do not make any particular assumptions about the heuristic used. We only require that the variable in σ_d is unassigned and occurs in the input program P (either as an atom or a body). Also note that $dlevel(\sigma_d)$ is set to the increment of dl in Line 16, so that σ_d is appended to the end of A in Line 17.

As an initial example, let us trace the computation of stable model $\{a, c\}$ of Program P_7 in Table 6.1.

$$P_7 = \left\{ \begin{array}{lll} a \leftarrow & c \leftarrow a, \sim d & e \leftarrow b, \sim f \\ b \leftarrow \sim a & d \leftarrow \sim c, \sim e & e \leftarrow e \end{array} \right\}$$

This and the following tables show the current assignment A at different stages of Algorithm 3. The leftmost column provides the value of the current decision level dl, whose corresponding decision literal is listed below σ_d. The only decision entry in Table 6.1 is $\mathbf{T}\{a, \sim d\}$. All other entries in A result from executing NogoodPropagation. That is, each entry under σ is unit-resulting for some nogood $\delta \in \Delta_{P_7} \cup \nabla$. The next two columns provide the specific group of nogoods comprising δ and the line of Algorithm 3 at which the assignment A and/or some nogood δ are inspected. Finally, the last column indicates the respective iteration of Algorithm 3's main loop. When applicable, we also indicate successful tests for violated nogoods and give the nogood δ recorded in ∇ along with the decision level dl to backjump to (thus abusing the column headed 'Origin' to indicate the jump's destination). For instance, such information is given in the last two lines of Table 6.2 below.

In fact, the stable model $\{a, c\}$ of P_7 is found and returned after two iterations of the main loop in Line 4 to 17 of Algorithm 3. The initial application of NogoodPropagation is identical to that of unit propagation in Figure 5.7.[1] Since the resulting assignment is still partial, we have to make a non-deterministic choice in Line 14, leading to the assignment of $\mathbf{T}$ to body $\{a, \sim d\}$. The second invocation of NogoodPropagation after re-entering the loop in Line 4 completes the assignment.

[1] See also the tableau derivation in Figure 5.6 on Page 81 as well as the iterated applications of operator Ω_{P_7} in (5.8) on Page 77.

		A		Origin	Line	Loop
dl	σ_d	σ	δ			

Table 6.1: Tracing the computation of stable model $\{a, c\}$ of P_7

dl	σ_d	σ	δ	Origin	Line	Loop
0		$\boldsymbol{F}f$	$\{\boldsymbol{F}f\}$	Δ_{P_7}	5	1
		$\boldsymbol{T}\varnothing$	$\{\boldsymbol{T}\varnothing\}$	Δ_{P_7}	5	
		$\underline{\boldsymbol{T}a}$	$\{\boldsymbol{F}a, \boldsymbol{T}\varnothing\}$	Δ_{P_7}	5	
		$\boldsymbol{F}\{\sim a\}$	$\{\boldsymbol{T}\{\sim a\}, \boldsymbol{T}a\}$	Δ_{P_7}	5	
		$\boldsymbol{F}b$	$\{\boldsymbol{T}b, \boldsymbol{F}\{\sim a\}\}$	Δ_{P_7}	5	
		$\boldsymbol{F}\{b, \sim f\}$	$\{\boldsymbol{T}\{b, \sim f\}, \boldsymbol{F}b\}$	Δ_{P_7}	5	
		$\boldsymbol{F}e$	$\{\boldsymbol{T}e, \boldsymbol{F}\{b, \sim f\}\}$	$\wedge_{P_7}$	5	
		$\boldsymbol{F}\{e\}$	$\{\boldsymbol{T}\{e\}, \boldsymbol{F}e\}$	Δ_{P_7}	5	
1	$\boldsymbol{T}\{a, \sim d\}$				16	
		$\boldsymbol{F}d$	$\{\boldsymbol{T}d, \boldsymbol{T}\{a, \sim d\}\}$	Δ_{P_7}	5	2
		$\underline{\boldsymbol{T}c}$	$\{\boldsymbol{F}c, \boldsymbol{T}\{a, \sim d\}\}$	Δ_{P_7}	5	
		$\boldsymbol{F}\{\sim c, \sim e\}$	$\{\boldsymbol{F}d, \boldsymbol{T}\{\sim c, \sim e\}\}$	Δ_{P_7}	5	

This is detected in Line 11 and the stable model is put out in Line 12. The entries characterizing stable model $\{a, c\}$ are underlined in Table 6.1.

Next, let us reconsider Program P_{24} from Page 72.

$$P_{24} = \left\{ \begin{array}{llll} a \leftarrow \sim b & c \leftarrow a, b & d \leftarrow a & e \leftarrow \sim a, \sim b \\ b \leftarrow \sim a & c \leftarrow d & d \leftarrow c, b & \end{array} \right\}$$

Although we have not yet detailed the subroutines of Algorithm 3, we give in Table 6.2 and 6.3 a full-fledged computation of P_{24}'s stable model $\{a, c, d\}$.

The first four iterations of the main loop of Algorithm 3 are illustrated in Table 6.2. At first, propagation yields no result and thus no consequences are drawn at decision level 0. Then, variable c is assigned $\boldsymbol{T}$ in Line 14. At the same time, the decision level is set to 1. Still no propagation takes place in the second iteration resulting in the non-deterministic consequence $\boldsymbol{F}\{\sim a, \sim b\}$. With the third iteration, NogoodPropagation extends the current assignment by $\boldsymbol{F}e$ to $(\boldsymbol{T}c, \boldsymbol{F}\{\sim a, \sim b\}, \boldsymbol{F}e)$. Given that this assignment is conflict-free yet partial, we select $\boldsymbol{F}\{\sim b\}$ and increment the decision level to 3. The fourth loop starts with NogoodPropagation extending A with eight deterministic consequences and adding loop nogood[2] $\{\boldsymbol{T}c, \boldsymbol{F}\{a\}, \boldsymbol{F}\{a, b\}\}$ to ∇. This nogood is contained in the current assignment and thus causes a conflict in Line 6. Given that our current decision level is well beyond 0, we let ConflictAnalysis transform the conflictual nogood into one being asserting at the backjump level (see Section 6.6 for details). As a result, we learn the nogood $\{\boldsymbol{T}c, \boldsymbol{F}a\}$ and backjump to level 1 by removing from the assignment all entries belonging to levels 2 and 3.

[2]This loop nogood is created by NogoodPropagation after detecting that $\{c, d\}$ has become an unfounded set (because its external support was invalidated; see Section 6.5 for details).

Table 6.2: Tracing the computation of stable model $\{a, c, d\}$ of P_{24}

dl	σ_d	A σ	δ	Origin	Line	Loop
1	Tc				16	1
2	$F\{\sim a, \sim b\}$				16	2
		Fe	$\{Te, F\{\sim a, \sim b\}\}$	$\Delta_{P_{24}}$	5	3
3	$F\{\sim b\}$				16	
		Fa	$\{Ta, F\{\sim b\}\}$	$\Delta_{P_{24}}$	5	4
		$F\{a\}$	$\{T\{a\}, Fa\}$	$\Delta_{P_{24}}$	5	
		$F\{a, b\}$	$\{T\{a, b\}, Fa\}$	$\Delta_{P_{24}}$	5	
		$T\{\sim a\}$	$\{F\{\sim a\}, Fa\}$	$\Delta_{P_{24}}$	5	
		Tb	$\{F\{\sim b\}, Fb\}$	$\Delta_{P_{24}}$	5	
		$T\{d\}$	$\{Tc, F\{a, b\}, F\{d\}\}$	$\Delta_{P_{24}}$	5	
		$T\{c, b\}$	$\{F\{c, b\}, Tc, Tb\}$	$\Delta_{P_{24}}$	5	
		Td	$\{Fd, T\{c, b\}\}$	$\Delta_{P_{24}}$	5	
			$\{Tc, F\{a\}, F\{a, b\}\}$	$\Lambda_{P_{24}}$	6	
			$\{Tc, Fa\}$	$dl=1$	8	

Upon entering our main loop for the fifth time, we have the assignment (Tc), decision level 1, and recorded the loop nogood $\{Tc, F\{a\}, F\{a, b\}\}$ as well as the conflict nogood $\{Tc, Fa\}$. This situation is similar to the one encountered at the beginning of iteration two; it is reflected by the first common lines in Table 6.2 and 6.3, respectively. That is, we are now back at decision level 1

Table 6.3: Tracing the computation of stable model $\{a, c, d\}$ of P_{24}

dl	σ_d	A σ	δ	Origin	Line	Loop
1	$\underline{Tc}$				16	1
		$\underline{Ta}$	$\{Tc, Fa\}$	∇	5	5
		$T\{a\}$	$\{F\{a\}, Ta\}$	$\Delta_{P_{24}}$	5	
		$F\{\sim a\}$	$\{T\{\sim a\}, Ta\}$	$\Delta_{P_{24}}$	5	
		$F\{\sim a, \sim b\}$	$\{T\{\sim a, \sim b\}, Ta\}$	$\Delta_{P_{24}}$	5	
		$\underline{Td}$	$\{Fd, T\{a\}\}$	$\Delta_{P_{24}}$	5	
		$T\{d\}$	$\{F\{d\}, Td\}$	$\Delta_{P_{24}}$	5	
		Fb	$\{Tb, F\{\sim a\}\}$	$\Delta_{P_{24}}$	5	
		$F\{a, b\}$	$\{T\{a, b\}, Fb\}$	$\Delta_{P_{24}}$	5	
		$F\{c, b\}$	$\{T\{c, b\}, Fb\}$	$\Delta_{P_{24}}$	5	
		Fe	$\{Te, F\{\sim a, \sim b\}\}$	$\Delta_{P_{24}}$	5	

and possess the same assignment as before. However, the situation has changed due to the recorded nogoods. While no propagation was possible in the second loop, ConflictAnalysis ensured that the learned conflict nogood is asserting after backjumping. Indeed, we see that unlike before, Nogood-Propagation now yields Ta from $\{Tc, Fa\}$. This is in turn completed to a total assignment. This assignment is detected in Line 11 and returned in Line 12 of Algorithm 3. Entries comprising true atoms are underlined in Table 6.3.

For a more complex example, reconsider Program P_{25} from Page 72.

$$P_{25} = \left\{ \begin{array}{llll} a \leftarrow \sim b & c \leftarrow a & d \leftarrow b, c & e \leftarrow b, \sim a \\ b \leftarrow \sim a & c \leftarrow b, d & d \leftarrow e & e \leftarrow c, d \end{array} \right\}$$

A computation of Algorithm 3 can proceed by successively picking decision entries Td, $F\{b, \sim a\}$, Tc, and $F\{\sim a\}$ at levels 1, 2, 3, and 4, respectively. Clearly, there is exactly one decision entry per level; each decision is immediately followed by a propagation step, performed before making the next decision. As we see in Table 6.4, propagation cannot derive any entry at decision levels 1 and 2.

Table 6.4: Tracing the computation of stable model $\{b, c, d, e\}$ of P_{25}

dl	A	δ		Origin	Line
1	Td				16
2	$F\{b, \sim a\}$				16
3	Tc				16
	$T\{c, d\}$	$\{F\{c, d\}, Tc, Td\}$		$\Delta_{P_{25}}$	5
	Te	$\{Fe, T\{c, d\}\}$		$\Delta_{P_{25}}$	5
	$T\{e\}$	$\{F\{e\}, Te\}$		$\Delta_{P_{25}}$	5
4	$F\{\sim a\}$				16
	Ta	$\{F\{\sim a\}, Fa\}$		$\Delta_{P_{25}}$	5
	$T\{a\}$	$\{F\{a\}, Ta\}$		$\Delta_{P_{25}}$	5
	$T\{\sim b\}$	$\{Ta, F\{\sim b\}\}$		$\Delta_{P_{25}}$	5
	Fb	$\{Tb, F\{\sim a\}\}$		$\Delta_{P_{25}}$	5
	$F\{b, c\}$	$\{T\{b, c\}, Fb\}$		$\Delta_{P_{25}}$	5
	$F\{b, d\}$	$\{T\{b, d\}, Fb\}$		$\Delta_{P_{25}}$	5
		$\{Td, F\{b, c\}, F\{b, \sim a\}\}$		$\Delta_{P_{25}}$	6
		$\{Td, F\{b, c\}, F\{b, \sim a\}\}$		$dl=2$	8

After the third decision, the entries shown immediately below decision entry Tc are unit-resulting for the respective nogoods δ in $\Delta_{P_{25}}$ with respect to A. Hence, they are added to A at decision level 3. Since A is still partial, decision entry $F\{\sim a\}$ is picked at level 4. The following propagation step yields a total assignment, which is actually a solution for $\Delta_{P_{25}}$. However, propagation also detects that the set $\{d, e\}$ is unfounded for P_{25} with respect to A. Hence, the corresponding loop nogoods

$\{Td, F\{b, c\}, F\{b, \sim a\}\}$ and $\{Te, F\{b, c\}, F\{b, \sim a\}\}$ are violated. Such violations are detected within NogoodPropagation and lead to the recording of some loop nogood from $\Lambda_{P_{25}}$ in ∇. In Table 6.4, we assume that $\{Td, F\{b, c\}, F\{b, \sim a\}\}$ is recorded, so that a conflict is encountered in Line 6 of Algorithm 3. Note that $F\{b, c\}$ is the single entry of this nogood assigned at decision level 4. Hence, $\{Td, F\{b, c\}, F\{b, \sim a\}\}$ is already asserting and directly returned by Conflict-Analysis in Line 8. The smallest decision level guaranteeing that the nogood is asserting is 2 because it is the maximum decision level among $dlevel(Td)$ and $dlevel(F\{b, \sim a\})$.[3]

Upon backjumping to decision level 2, all entries added to A at levels 3 and 4 are removed, and only entries assigned at levels 1 and 2 are retained. This situation is reflected by the first two lines of Table 6.4 and 6.5, representing assignment $(Td, F\{b, \sim a\})$. In analogy to the previous

Table 6.5: Tracing the computation of stable model $\{b, c, d, e\}$ of P_{25}					
dl	A	δ		Origin	Line
1	Td				16
2	$F\{b, \sim a\}$				16
	$T\{b, c\}$	$\{Td, F\{b, c\}, F\{b, \sim a\}\}$		∇	5
	Tb	$\{T\{b, c\}, Fb\}$		$\Delta_{P_{25}}$	5
	Ta	$\{F\{b, \sim a\}, Tb, Fa\}$		$\Delta_{P_{25}}$	5
	$T\{\sim a\}$	$\{Tb, F\{\sim a\}\}$		$\Delta_{P_{25}}$	5
		$\{T\{\sim a\}, Ta\}$		$\Delta_{P_{25}}$	6
		$\{F\{b, \sim a\}, Td\}$		$dl=1$	8

example, the asserting nogood $\{Td, F\{b, c\}, F\{b, \sim a\}\}$ in ∇ enables the derivation of further entries by unit propagation. This ultimately results in another conflict. However, this time the completion nogood $\{T\{\sim a\}, Ta\}$ is violated. Starting from it, ConflictAnalysis determines the asserting nogood $\{F\{b, \sim a\}, Td\}$. As a consequence, Algorithm 3 returns to decision level 1, as illustrated in Table 6.6. In contrast to Table 6.4, $T\{b, \sim a\}$ is now unit-resulting for $\{F\{b, \sim a\}, Td\}$. A final propagation step leads to a total assignment not violating any nogood in $\Delta_{P_{25}} \cup \nabla$. (Notably, the nogoods in $\Lambda_{P_{25}}$ are left implicit and merely tested within NogoodPropagation via an unfounded set checking subroutine; see Section 6.5.) The associated stable model of P_{25}, $\{b, c, d, e\}$, is returned as the result. The corresponding entries are underlined in Table 6.6.

6.4   NOGOOD PROPAGATION

The subroutine NogoodPropagation for deterministically extending an assignment combines unit propagation on completion nogoods in Δ_P and recorded nogoods in ∇ (Lines 3–9 of Algorithm 4)

---

[3]The peculiarity that ConflictAnalysis may be launched with an asserting (loop) nogood results from the "unidirectional" propagation of loop nogoods in current ASP solvers: ASP solvers implement tableau rule *WFN*, but not its contrapositive *WFJ*, although both tableau rules are logically based on loop nogoods.

Table 6.6: Tracing the computation of stable model $\{b, c, d, e\}$ of P_{25}

dl	A	δ	Origin	Line
1	Td			16
	$T\{b, \sim a\}$	$\{F\{b, \sim a\}, Td\}$	∇	5
	Tb	$\{T\{b, \sim a\}, Fb\}$	$\Delta_{P_{25}}$	5
	Fa	$\{T\{b, \sim a\}, Ta\}$	$\Delta_{P_{25}}$	5
	$T\{\sim a\}$	$\{Tb, F\{\sim a\}\}$	$\Delta_{P_{25}}$	5
	$F\{\sim b\}$	$\{T\{\sim b\}, Tb\}$	$\Delta_{P_{25}}$	5
	$F\{a\}$	$\{T\{a\}, Fa\}$	$\Delta_{P_{25}}$	5
	Te	$\{Fe, T\{b, \sim a\}\}$	$\Delta_{P_{25}}$	5
	$T\{e\}$	$\{F\{e\}, Te\}$	$\Delta_{P_{25}}$	5
	$T\{b, d\}$	$\{F\{b, d\}, Tb, Td\}$	$\Delta_{P_{25}}$	5
	Tc	$\{Fc, T\{b, d\}\}$	$\Delta_{P_{25}}$	5
	$T\{b, c\}$	$\{F\{b, c\}, Tb, Tc\}$	$\Delta_{P_{25}}$	5
	$T\{c, d\}$	$\{F\{c, d\}, Tc, Td\}$	$\Delta_{P_{25}}$	5

with unfounded set checking (Lines 10–15). While unit propagation is always run to a fixpoint (or a conflict), unfounded set checks are only done when the input program P is not tight.[4]

Note that the construction of Σ (in Line 5) guarantees that variables are unassigned before a corresponding entry is inserted in Line 8. Along with the way $dlevel(\sigma)$ is set in Line 7, any assignment A computed by NogoodPropagation is ordered and all newly assigned entries σ are implied by $\Delta_P \cup \nabla$ with respect to A.

The idea of integrating unfounded set checking with unit propagation is to trigger the consecutive falsification of unfounded atoms by recording loop nogoods from Λ_P in ∇. To see this, consider Normal program P_{27} along with assignment $A = (Tb, T\{\sim a\}, Fa, F\{\sim b\})$.

$$P_{27} = \left\{ \begin{array}{lll} a \leftarrow \sim b & c \leftarrow \sim b & d \leftarrow c \\ b \leftarrow \sim a & c \leftarrow d, e & e \leftarrow d \end{array} \right\}$$

We observe that $U = \{c, d, e\}$ is an unfounded set for P_{27} with respect to A. Therefore, given that $EB_{P_{27}}(U) \subseteq A^F$, all literals Fc, Fd, and Fe are unit-resulting for the three loop nogoods $\{Tc, F\{\sim b\}\}$, $\{Td, F\{\sim b\}\}$, and $\{Te, F\{\sim b\}\}$ in $\Lambda_{P_{27}}$. While neither Fc, Fd, nor Fe are unit-resulting for a completion nogood in $\Delta_{P_{27}}$, all of them (along with $F\{c\}$, $F\{d\}$, and $F\{d, e\}$) are obtained by unit propagation from $\Delta_{P_{27}} \cup \{\{Tc, F\{\sim b\}\}\}$. That is, the addition of a single loop nogood from $\Lambda_{P_{27}}$ to ∇ may falsify the entire unfounded set by unit propagation. However, whether a single loop nogood is sufficient to falsify a whole unfounded set depends on the program's structure. For instance, extending P_{27} with $d \leftarrow e$ to P'_{27} inhibits the derivation of Fd and Fe by

[4]Otherwise, all unfounded sets U are already falsified, that is, $U \subseteq A^F$.

Algorithm 4: NOGOODPROPAGATION

    Input : A normal program P, a set ∇ of nogoods, and an assignment A.
    Output: An extended assignment and set of nogoods.

1   $U := \emptyset$ *// unfounded set*

2 **loop**

3   **repeat**

4   **if** $\delta \subseteq A$ **for some** $\delta \in \Delta_P \cup \nabla$ **then return** (A, ∇) *// conflict*

5          $\Sigma := \{\delta \in \Delta_P \cup \nabla \mid \delta \setminus A = \{\overline{\sigma}\}, \sigma \notin A\}$ *// unit-resulting nogoods*

6   **if** $\Sigma \neq \emptyset$ **then let** $\overline{\sigma} \in \delta \setminus A$ **for some** $\delta \in \Sigma$ **in**

7              $dlevel(\sigma) := \max(\{dlevel(\rho) \mid \rho \in \delta \setminus \{\overline{\sigma}\}\} \cup \{0\})$

8              $A := A \circ \sigma$

9   **until** $\Sigma = \emptyset$

10   **if** $loop(P) = \emptyset$ **then return** (A, ∇)

11      $U := U \setminus A^{F}$

12   **if** $U = \emptyset$ **then** $U := \mathrm{UnfoundedSet}(P, A)$

13   **if** $U = \emptyset$ **then return** (A, ∇) *// no unfounded set* $\emptyset \subset U \subseteq atom(P) \setminus A^{F}$

14   **let** $a \in U$ **in**

15          $\nabla := \nabla \cup \{\{\boldsymbol{T}a\} \cup \{\boldsymbol{F}B \mid B \in EB_P(U)\}\}$ *// record loop nogood*

---

unit propagation because the (circular) support among d and e is not eliminated by falsifying c. Although we still derive $\boldsymbol{F}\{c\}$, that is, the rule $d \leftarrow c$ becomes inapplicable, so that $EB_{P'_{27}}(\{d, e\}) \subseteq (A \cup \{\boldsymbol{F}c, \boldsymbol{F}\{c\}\})^{F}$, the nogoods in $\Delta_{P'_{27}} \cup \{\{\boldsymbol{T}c, \boldsymbol{F}\{\sim b\}\}\}$ do not suffice to imply $\boldsymbol{F}d$ and $\boldsymbol{F}e$. This shows that $U \setminus (A \cup \{\boldsymbol{F}c, \boldsymbol{F}\{c\}\})^{F} = \{c, d, e\} \setminus \{c\} = \{d, e\}$ remains as a smaller unfounded set.

The observation made in the previous example motivates the strategy of Algorithm 4 to consecutively falsify the elements of an unfounded set U. At the start, no (non-empty) unfounded set has been determined, and so U is initialized to be empty in Line 1. Provided that unit propagation in Lines 3–9 finishes without conflict and that P is not tight, we remove all false atoms from U in Line 11. In the first iteration of the outer loop, U stays empty, and our subroutine for unfounded set detection is queried in Line 12. The crucial assumption is that $\mathrm{UnfoundedSet}(P, A)$ returns an unfounded set $U \subseteq atom(P) \setminus A^{F}$ such that U is non-empty whenever some non-empty subset of $atom(P) \setminus A^{F}$ is unfounded. Then, if a non-empty U is returned, we have $EB_P(U) \subseteq A^{F}$. Hence, the addition of a loop nogood $\{\boldsymbol{T}a\} \cup \{\boldsymbol{F}B \mid B \in EB_P(U)\}$ to ∇ for an arbitrary $a \in U$ (in Line 15) yields either a conflict (because $\boldsymbol{T}a \in A$) or $\boldsymbol{F}a$ upon unit propagation. In the latter case, further elements of U may be falsified by unit propagation in Lines 3–9. When we afterward reconsider the previously determined unfounded set U, the removal of false atoms in Line 11 is guaranteed to result in another (smaller) unfounded set $U \setminus A^F$. Hence, if $U \setminus A^F$ is non-empty (checked in Line 12 before computing any new unfounded set), NogoodPropagation proceeds by adding the next loop nogood to ∇, which as before yields either a conflict or a unit-resulting entry. In this way, once a non-empty unfounded set U has been detected, it is falsified element by element. Only after expending all elements of U, a new unfounded set is computed.

All in all, NogoodPropagation terminates as soon as a conflict is encountered (in Line 4) or with an assignment free of any (non-empty) unfounded subsets of $atom(P) \setminus A^F$. If P is tight, the latter is immediately verified in Line 10. Otherwise, the subroutine UnfoundedSet failed to detect a non-empty unfounded set (of non-false atoms) before finishing in Line 13.

To illustrate how NogoodPropagation utilizes nogoods, reconsider the computation of CDNL-ASP shown in Table 6.4. All implied entries are unit-resulting for nogoods in $\Delta_{P_{25}} \cup \nabla$ and successively derived by unit propagation. In particular at decision level 4, the implied entries σ have antecedents δ in $\Delta_{P_{25}}$. All entries of each δ except for $\overline{\sigma}$ are already contained in A when σ is inserted into A. The impact of loop nogoods in $\Lambda_{P_{25}}$ can be observed on the conflict encountered at decision level 4. Here, we have that $U = \{d, e\} \subseteq A^T$ is unfounded, so that A violates each of the loop nogoods $\{Td, F\{b, c\}, F\{b, \sim a\}\}$ and $\{Te, F\{b, c\}, F\{b, \sim a\}\}$. After detecting the unfounded set U and recording $\{Td, F\{b, c\}, F\{b, \sim a\}\}$ in ∇, its violation gives rise to leaving Algorithm 4 in Line 4.

To sum up, NogoodPropagation interleaves unit propagation with the recording of loop nogoods. The latter is done only if the input program is not tight and if the falsity of unfounded atoms cannot be derived by unit propagation via other available nogoods. In this way, the approach clearly favors unit propagation over unfounded set computation. For one thing, unit propagation does not contribute new dynamic nogoods to ∇, so that it is more "economic" than unfounded set checking. For another, although unfounded set detection algorithms are of linear time complexity, they need to analyze a logic program in a global fashion and may get stuck half-way, inspecting substantial program parts without eventually detecting any unfounded sets of interest. In contrast to this, unit propagation stays local by investigating only nogoods related to literals that become assigned. Hence, the effort made to identify unit-resulting literals is much less than unfounded set checking. But given that unfounded set checking is mandatory for soundness (with respect to total assignments) and also helps to detect inherent conflicts early (with respect to partial assignments), the subroutine described next is nonetheless an integral part of NogoodPropagation.

6.5   UNFOUNDED SET CHECKING

The subroutine for unfounded set detection is invoked on non-tight programs once unit propagation reaches a fixpoint without any conflict or remaining unfalsified unfounded atoms. As a matter of fact, a fixpoint of unit propagation allows us to focus on unfounded sets of non-false atoms contained in non-trivial strongly connected components of a program's dependency graph.

Given a program P along with its positive dependency graph $G(P)$, we define for each $a \in atom(P)$ the set $scc(a)$ as being composed of all atoms belonging to the same strongly connected component as a in $G(P)$. We call atom a *cyclic*, if its strongly connected component in $G(P)$ is non-trivial, and *acyclic* otherwise. In other words, a is cyclic when there is some rule $r \in P$ such that $head(r) \in scc(a)$ and $body(r)^+ \cap scc(a) \neq \emptyset$. In fact, unfounded set checking can concentrate on cyclic atoms, since only they can belong to (unfounded) loops.

Beyond static information about strongly connected components, our UnfoundedSet algorithm makes use of *source pointers* to indicate non-circular supports of atoms. The idea is to associate every cyclic atom $a \in atom(P)$ either with one of its rule bodies in $body_P(a)$ pointing itself to a chain of rule bodies witnessing that a cannot be unfounded, or with one of the special-purpose symbols $\circ$ and $\times$. The associated source pointer of a is denoted by $source(a)$. Thus, as long as $source(a)$ remains "intact," a can be ignored by unfounded set checks. In this way, source pointers enable lazy, incremental unfounded set checking relative to recent changes of an assignment.

For an appropriate initialization, we define the initial source pointer configuration for a program P as follows.

$$source(a) \quad = \quad \begin{cases} \circ & \text{if } a \in atom(P) \text{ is cyclic} \\ \times & \text{if } a \in atom(P) \text{ is acyclic} \end{cases}$$

While $\times$ expresses that an acyclic atom a does not need to be linked to any element of $body_P(a)$, $\circ$ indicates that a non-circular support for a cyclic atom a still needs to be determined. We assume that the initial source pointer configuration for P is in place when invoking CDNL-ASP(P).

Given a program P and an assignment A, Algorithm 5 starts in Line 1 by collecting non-false cyclic atoms a whose source pointers are either false ($source(a) \in A^F$) or yet undetermined ($source(a) = \circ$). The resulting set S delineates the initial *scope* of all atoms whose non-circular support is in question. In the following Lines 2 to 5, the scope is successively extended by adding atoms whose source pointers (positively) depend on it.[5] In fact, the loop in Lines 6–17 aims at reestablishing source pointers for the atoms in S via rules whose bodies do not (positively) rely on S. If successful, these rules provide a non-circular support for the atoms in question. Otherwise, an unfounded set is detected whenever source pointers cannot be reestablished.

Let us describe the second part of Algorithm 5 in more detail. As long as the scope S is non-empty, some atom $a \in S$ is picked in Line 6 as a starting point for the construction of a non-empty unfounded set U. If $EB_P(U) \subseteq A^F$ is found to hold in Line 9, the unfounded set U is immediately returned. In this way, NogoodPropagation can in turn falsify the atoms in U by unit propagation. Otherwise, some external body $B \in EB_P(U) \setminus A^F$ is selected in Line 10 for inspection. If B^+ shares atoms with the scope S belonging to the same strongly connected component as the starting point a (checked in Line 11), we add these common atoms to U in Line 16. As a side-effect, this eliminates B from the external bodies of the augmented set U. On the other hand, if such atoms do not exist in B^+, the rule body B can non-circularly support all of its associated head atoms $c \in U$. Then, in

---

[5]For this, define $source(a) = \emptyset$ whenever $source(a) \notin P$, that is, if $source(a) = \circ$ or $source(a) = \times$.

Algorithm 5: UNFOUNDEDSET

    Input   : A normal program P and an assignment A.
    Output: An unfounded set of P with respect to A.
    Global : Source pointers $\{source(a) \mid a \in atom(P)\}$.

1 $S := \{a \in atom(P) \setminus A^F \mid source(a) \in A^F \cup \{\circ\}\}$ *// initialize scope S*

2 **repeat**

3     $T := \{a \in atom(P) \setminus (A^F \cup S) \mid source(a)^+ \cap (scc(a) \cap S) \neq \emptyset\}$

4     $S := S \cup T$ *// extend scope S*

5 **until** $T = \emptyset$

6 **while** $S \neq \emptyset$ **do let** $a \in S$ **in**   *// select starting point a*

7     $U := \{a\}$

8   **repeat**

9   **if** $EB_P(U) \subseteq A^F$ **then return** U *// (non-empty) unfounded set*

10   **let** $B \in EB_P(U) \setminus A^F$ **in**

11   **if** $B^+ \cap (scc(a) \cap S) = \emptyset$ **then**   *// shrink U*

12   **foreach** $c \in U$ such that $B \in body_P(c)$ **do**

13             $source(c) := B$

14             $U := U \setminus \{c\}$

15             $S := S \setminus \{c\}$

16   **else** $U := U \cup (B^+ \cap (scc(a) \cap S))$ *// extend U*

17   **until** $U = \emptyset$

18 **return** $\emptyset$ *// no unfounded set $\emptyset \subset U \subseteq atom(P) \setminus A^F$*

Lines 12–15, the source pointers of all such atoms c are consecutively set to B. And all such atoms c are removed from both the unfounded set U under construction and the scope S. This loop continues until either U becomes empty (in Line 17) or a (non-empty) unfounded set U is detected and returned in Line 9. In the former case, the remaining atoms of S are investigated by recommencing the outer loop. Finally, if the scope S runs empty, source pointers could be reestablished for all atoms contained in S, and UnfoundedSet returns the empty unfounded set in Line 18.

    In order to provide further insight, let us stress some major design principles underlying Algorithm UnfoundedSet.

1. At each stage of the loop in Lines 6–17, all atoms of U belong to $scc(a)$, where a is the atom added first to U in Line 7. This is because further atoms, added to U in Line 16, are elements of $scc(a)$.

However, $U \subseteq scc(a)$ does not necessarily imply $a \in U$ for a (non-empty) unfounded set U returned in Line 9.

2. At each stage of the loop in Lines 6–17, we have that $U \subseteq S$ since all atoms added to U in either Line 7 or 16 belong to S. Hence, we have $c \in S$ whenever $source(c)$ is set to an external body $B \in body_P(c)$ in Line 13. Unlike this, we have $B^+ \cap (scc(a) \cap S) = \emptyset$ as has been checked before in Line 11. This makes sure that setting $source(c)$ to B does not introduce any cycles via source pointers.

3. Once detected, a (non-empty) unfounded set U is immediately returned in Line 9, and NogoodPropagation takes care of falsifying all elements of U before checking for any further unfounded set. This reduces overlaps with unit propagation on the completion nogoods in Δ_P, because it already handles singleton unfounded sets (and bodies relying on them).[6]

4. After establishing the initial source pointer configuration for a program, source pointers are only set in Line 13 when reestablishing a potential non-circular support for an atom c. In fact, the source pointer of an atom c in an unfounded set U returned in Line 9 needs not and is not reset to $\circ$. Rather, we admit $source(c) \in A^F$ as long as $c \in A^F$ (derived within Nogood-Propagation upon falsifying U). Thus, it is only when c becomes unassigned later on (after backjumping), that its source pointer $source(c)$ is reconsidered in Line 1. This amounts to lazy unfounded set checking.

Let us illustrate Algorithm 5 on some invocations of UnfoundedSet(P_{25}, A) made when computing the stable model $\{b, c, d, e\}$ of Program P_{25}. To this end, we indicate in Table 6.7 internal states of UnfoundedSet(P_{25}, A) when queried with P_{25} and fixpoints A of unit propagation at decision levels 0, 2, and 4 in Table 6.4, respectively. Looking at the dependency graph $G(P_{25})$ in Figure 5.3 on Page 73, we observe that $scc(c) = scc(d) = scc(e) = \{c, d, e\}$ while a and b are acyclic. Hence, before the first launch of UnfoundedSet(P_{25}, A) at decision level 0, we have to install the initial source pointer configuration for P_{25}:

$$
\begin{aligned}
source(a) \;=\; source(b) \;&=\; \times \qquad \text{and} \\
source(c) \;=\; source(d) \;=\; source(e) \;&=\; \circ \;\; .
\end{aligned}
$$

According to Line 1 in Algorithm 5, we obtain the scope $S = \{c, d, e\}$. Next, let us choose atom e in Line 6 and add it to U in Line 7. Following this, we select $\{c, d\} \in EB_{P_{25}}(\{e\})$ in Line 10. The fact that $\{c, d\} \cap (scc(e) \cap S)$ equals $\{c, d\}$ makes us augment U with both c and d in Line 16, resulting in an intermediate state where $U = \{c, d, e\}$. Let us select next $\{b, \sim a\} \in EB_{P_{25}}(\{c, d, e\})$ in Line 10, for which we get $\{b\} \cap (scc(e) \cap S) = \emptyset$ in Line 11. Hence, $source(e)$ is set to $\{b, \sim a\}$ in Line 13, and e is removed from U and S in Lines 14 and 15, respectively. In the same manner, the source pointers of d and c are set in the following iterations of the loop in Lines 8–17 to $\{e\}$ and $\{b, d\}$, respectively. Afterward, we have that $U = S = \emptyset$, so that the empty unfounded set (surrounded

---

[6]This amounts to inferences drawn by tableau rule *FFA* and *BFA*.

				Table 6.7: Trace of UNFOUNDEDSET(P_{25}, A)	
dl	$source(p)$	S	U	$B \in EB_{P_{25}}(U) \setminus A^F$	Line
0		$\{c,d,e\}$			1
		$\{c,d,e\}$	$\{e\}$		7
		$\{c,d,e\}$	$\{c,d,e\}$	$\{c,d\}$	16
	$source(e)$	$\{c,d\}$	$\{c,d\}$	$\{b,\sim a\}$	13
	$source(d)$	$\{c\}$	$\{c\}$	$\{e\}$	13
	$source(c)$	$\emptyset$	$\boxed{\emptyset}$	$\{b,d\}$	13
2	$\boldsymbol{F}\{b,\sim a\}$	$\{e\}$			1
	$\{e\}$	$\{d,e\}$			4
	$\{b,d\}$	$\{c,d,e\}$			4
		$\{c,d,e\}$	$\{d\}$		7
		$\{c,d,e\}$	$\{c,d\}$	$\{b,c\}$	16
	$source(c)$	$\{d,e\}$	$\{d\}$	$\{a\}$	13
	$source(d)$	$\{e\}$	$\emptyset$	$\{b,c\}$	13
		$\{e\}$	$\{e\}$		7
	$source(e)$	$\emptyset$	$\boxed{\emptyset}$	$\{c,d\}$	13
4	$\boldsymbol{F}\{b,c\}$	$\{d\}$			1
	$\{c,d\}$	$\{d,e\}$			4
		$\{d,e\}$	$\{e\}$		7
		$\{d,e\}$	$\boxed{\{d,e\}}$	$\{c,d\}$	16

by a box in Table 6.7) is returned in Line 18. Given that there is no non-empty unfounded set, no entry is derived by unit propagation at decision level 0, as also indicated by omitting this level in Table 6.4. Despite this, the execution of UnfoundedSet(P_{25}, A) has led to the following source pointer configuration for P_{25}:

$$
\begin{aligned}
source(a) \;&=\; source(b) \;=\; \times, \\
source(c) \;&=\; \{b,d\}, \\
source(d) \;&=\; \{e\}, \qquad \text{and} \\
source(e) \;&=\; \{b,\sim a\}.
\end{aligned}
$$

We refrain from describing the invocation of UnfoundedSet(P_{25}, $(\boldsymbol{T}d)$) at decision level 1 in Table 6.7 because it yields an empty scope.

Unlike this, the call of UnfoundedSet(P_{25}, $(\boldsymbol{T}d, \boldsymbol{F}\{b,\sim a\})$) at decision level 2 is faced with a falsified source pointer. We have that $source(e) = \{b,\sim a\} \in A^F$, so that $S = \{e\}$ is obtained in Line 1 of Algorithm 5. In Lines 2–5, we successively add d and c to S because $source(d)^+ \cap S = \{e\} \cap \{e\} \neq \emptyset$ and $source(c)^+ \cap (S \cup \{d\}) = \{b,d\} \cap \{d,e\} \neq \emptyset$. Afterward, let us first add d to U in Line 7 and select $\{b, c\} \in EB_{P_{25}}(\{d\})$ in Line 10, leading to $U = \{d\} \cup (\{b, c\} \cap (scc(d) \cap S)) = \{c, d\}$. When investigating $\{a\} \in EB_{P_{25}}(\{c, d\})$ and again $\{b, c\} \in EB_{P_{25}}(\{d\})$ in the next two iterations of the loop in Lines 8–17, we set $source(c)$ to $\{a\}$ and $source(d)$ to $\{b, c\}$, respectively, resulting in $U = \emptyset$ and $S = \{e\}$. Since $S \neq \emptyset$, another iteration of the loop in Lines 6–17 adds e to U and then removes it from U and S along with setting $source(e)$ to $\{c, d\}$. Given $U = S = \emptyset$, we get the empty unfounded set (again surrounded by a box) as result in Line 18. As well, we get the following source pointer configuration for P_{25}:

$$
\begin{aligned}
source(a) \;=\; source(b) \;&=\; \times, \\
source(c) \;&=\; \{a\}, \\
source(d) \;&=\; \{b, c\}, \qquad \text{and} \\
source(e) \;&=\; \{c, d\}.
\end{aligned}
$$

At decision level 3, unfounded set checking is again without effect because no rule body and, in particular, no source pointer is falsified.

However, at decision level 4, we have that $source(d) = \{b, c\} \in A^{\mathbf{F}}$, and thus we get $S = \{d\}$ in Line 1 of Algorithm 5. In an iteration of the loop in Lines 2–5, we further add e to S because $source(e)^{+} \cap S = \{c, d\} \cap \{d\} \neq \emptyset$, while c stays unaffected in view of $source(c) = \{a\} \notin A^{\mathbf{F}}$. After adding e to U in Line 7, U is further extended to $\{d, e\}$ in Line 16, given that $\{c, d\} \in EB_{P_{25}}(\{e\})$ and $\{c, d\} \cap (scc(e) \cap S) = \{d\}$. We have now obtained $U = \{d, e\}$, and we get that $EB_{P_{25}}(\{d, e\}) = \{\{b, c\}, \{b, \sim a\}\} \subseteq A^{\mathbf{F}}$. That is, the termination condition in Line 9 applies, and Unfounded-Set(P_{25}, A) returns the (non-empty) unfounded set $\{d, e\}$ (and leaves the above source pointer configuration intact).

To conclude the example, we observe in Table 6.5 that adding the loop nogood $\{\mathbf{T}d, \mathbf{F}\{b, c\}, \mathbf{F}\{b, \sim a\}\}$ to ∇ leads to a conflict at decision level 4. After backjumping to decision level 2, NogoodPropagation encounters a conflict before launching UnfoundedSet(P_{25}, A). Hence, UnfoundedSet(P_{25}, A) is only queried again with respect to the total assignment A derived by unit propagation after returning to decision level 1. In view of $source(c) = \{a\} \in A^{\mathbf{F}}$, this final invocation (not shown in Table 6.7) makes us reset source pointers to the configuration at decision level 0:

$$
\begin{aligned}
source(a) \;=\; source(b) \;&=\; \times, \\
source(c) \;&=\; \{b, d\}, \\
source(d) \;&=\; \{e\}, \qquad \text{and} \\
source(e) \;&=\; \{b, \sim a\}.
\end{aligned}
$$

As this yields only the empty unfounded set (of non-false atoms), NogoodPropagation terminates without conflict, and CDNL-ASP(P_{25}) returns the stable model $\{b, c, d, e\}$.

6.6 CONFLICT ANALYSIS

The purpose of the subroutine for conflict analysis is to determine an asserting nogood, ensuring that one of its entries is unit-resulting after backjumping. To this end, a violated nogood $\delta \subseteq A$

Algorithm 6: CONFLICTANALYSIS

Input   : A non-empty violated nogood δ, a normal program P, a set ∇ of nogoods, and an assignment A.

Output: A derived nogood and a decision level.

1 **loop**
2 $\quad$ **let** $\sigma \in \delta$ **such that** $\delta \setminus A[\sigma] = \{\sigma\}$ **in**
3 $\quad\quad$ $k := \max(\{dlevel(\rho) \mid \rho \in \delta \setminus \{\sigma\}\} \cup \{0\})$
4 $\quad\quad$ **if** $k = dlevel(\sigma)$ **then**
5 $\quad\quad\quad$ **let** $\varepsilon \in \Delta_P \cup \nabla$ **such that** $\varepsilon \setminus A[\sigma] = \{\overline{\sigma}\}$ **in**
6 $\quad\quad\quad\quad$ $\delta := (\delta \setminus \{\sigma\}) \cup (\varepsilon \setminus \{\overline{\sigma}\})$ $\qquad$ *// resolution*
7 $\quad\quad$ **else return** (δ, k)

---

is resolved against antecedents ε of implied entries $\sigma \in \delta$ for obtaining a new violated nogood $(\delta \setminus \{\sigma\}) \cup (\varepsilon \setminus \{\overline{\sigma}\})$. Iterated resolution proceeds in inverse order of assigned entries, resolving first over the entry $\sigma \in \delta$ assigned last in A, and stops as soon as δ contains exactly one entry assigned at the decision level where the conflict is encountered. This entry is called the *unique implication point* (UIP). The strategy of stopping once the first UIP is found is referred to as *First-UIP* scheme.

ConflictAnalysis is performed by Algorithm 6 according to the First-UIP scheme. In fact, the loop in Lines 1–7 proceeds by resolving over the entry σ of the violated nogood δ assigned last in A (given that $\delta \setminus A[\sigma] = \{\sigma\}$ is required in Line 2) until the *assertion level*, that is, the greatest level $dlevel(\rho)$ associated with entries $\rho \in \delta \setminus \{\sigma\}$, is different from (and actually smaller than) $dlevel(\sigma)$. If so, nogood δ and assertion level k (determined in Line 3) are returned in Line 7. Since $\delta \subseteq A$, we have that $\overline{\sigma}$ is unit-resulting for δ after backjumping to decision level k. Otherwise, if $k = dlevel(\sigma)$, σ is an implied entry, so that some antecedent $\varepsilon \in \Delta_P \cup \nabla$ of σ can be chosen in Line 5 and used for resolution against δ in Line 6. Note that there may be several antecedents of σ in $\Delta_P \cup \nabla$, and thus the choice of ε in Line 5 is, in general, non-deterministic. Regarding the termination of Algorithm 6, note that a decision entry σ_d (cf. Algorithm 3) is the first entry in A at level $dlevel(\sigma_d)$, and σ_d is also the only entry at $dlevel(\sigma_d)$ that is not implied. Given that ConflictAnalysis is only applied to nogoods violated at decision levels beyond 0, all conflict resolution steps are well-defined and stop at the latest at a decision entry σ_d. However, resolving up to σ_d can be regarded as the worst case, given that the First-UIP scheme aims at few resolution steps to obtain a nogood that is "close" to a conflict at hand.

To illustrate Algorithm 6, let us inspect the resolution steps shown in Table 6.8. They are applied when resolving the violated nogood $\{T\{\sim a\}, Ta\}$ against the antecedents shown in Table 6.5. This is done upon analyzing the conflict encountered at decision level 2. In Table 6.8, we box both the entry σ of δ assigned last in A as well as its complement $\overline{\sigma}$ in an antecedent ε of σ; further entries

Table 6.8: CONFLICTANALYSIS at decision level 2 in Table 6.5	
δ	ε
$\left\{\boxed{T\{\sim a\}}, \underline{Ta}\right\}$	$\left\{\underline{Tb}, \boxed{F\{\sim a\}}\right\}$
$\left\{\boxed{Ta}, \underline{Tb}\right\}$	$\left\{F\{b, \sim a\}, \underline{Tb}, \boxed{Fa}\right\}$
$\left\{\boxed{Tb}, F\{b, \sim a\}\right\}$	$\left\{T\{b, c\}, \boxed{Fb}\right\}$
$\left\{\boxed{T\{b, c\}}, \underline{F\{b, \sim a\}}\right\}$	$\left\{Td, \boxed{F\{b, c\}}, \underline{F\{b, \sim a\}}\right\}$
$\left\{\boxed{F\{b, \sim a\}}, Td\right\}$	

assigned at decision level 2 are underlined. The final result of iterated resolution, $\{F\{b, \sim a\}, Td\}$, contains $F\{b, \sim a\}$ as the single entry assigned at decision level 2. Unlike this, the entry Td has been assigned at level 1. In this example, the (first) UIP, $F\{b, \sim a\}$, happens to be the decision entry at level 2.

In general, however, a first UIP is not necessarily a decision entry. For instance in Table 6.4, the UIP $F\{b, c\}$ in the asserting nogood $\{Td, F\{b, c\}, F\{b, \sim a\}\}$ is an implied entry. Also recall that $\{Td, F\{b, c\}, F\{b, \sim a\}\}$ served as the starting point for ConflictAnalysis, containing a (first) UIP without requiring any resolution step. This phenomenon is due to "unidirectional" propagation of loop nogoods, given that unfounded set checks merely identify unfounded atoms, but not rule bodies that must necessarily hold for (non-circularly) supporting some true atom. In Table 6.4, the fact that $T\{b, c\}$ is required from decision level 2 on is only recognized at level 4, where assigning $F\{b, c\}$ leads to a conflict. In view of this, Algorithm 5 can be understood as a checking routine guaranteeing the soundness of CDNL-ASP, while its inference capabilities do not match (full) unit propagation on loop nogoods.[7]

Despite the fact that conflict resolution in ASP can be done in the same fashion as in SAT, the input format of logic programs makes it less predetermined. For one thing, the completion nogoods in Δ_P contain rule bodies as structural variables for the sake of a compact representation. For another, the number of (relevant) inherent loop nogoods in Λ_P may be exponential. Fortunately, the satisfaction of Λ_P can be checked in linear time (via Algorithm 5), so that an explicit representation of its elements is not required. However, NogoodPropagation records loop nogoods from Λ_P that are antecedents to make them easily accessible in ConflictAnalysis.

---

[7] In other words, current ASP solvers implement tableau rule *WFN* but not its contrapositive *WFJ*, although both tableau rules are logically based on loop nogoods.

6.7 REFERENCES AND FURTHER READING

The success story of ASP has its roots in the early availability of ASP solvers, beginning with the *smodels* system (Simons et al., 2002), followed by *dlv* (Leone et al., 2006), SAT-based ASP solvers, like *assat* (Lin and Zhao, 2004) and *cmodels* (Giunchiglia et al., 2006), and more recently the conflict-driven learning ASP solver *clasp* (Gebser et al., 2007a). While traditional ASP solvers like *smodels* and *dlv* rely on DPLL-based algorithms, others build either directly or indirectly (via SAT solvers) on CDCL-based methods.

The Davis-Putman-Logemann-Loveland (DPLL) procedure was introduced in Davis and Putman (1960), Davis et al. (1962). The first principles of Conflict-Driven Clause Learning (CDCL) were gradually developed by Bayardo and Schrag (1997), Marques-Silva and Sakallah (1999), Moskewicz et al. (2001), and Zhang et al. (2001). Eén and Sörensson (2004) present an excellent introduction to algorithms and data structures for implementing CDCL. See also Darwiche and Pipatsrisawat (2009), Marques-Silva et al. (2009). Both DPLL and CDCL have firm proof-theoretic foundations in propositional resolution (Beame and Pitassi, 1998, Beame et al., 2004, Pipatsrisawat and Darwiche, 2011). While DPLL amounts to a restricted form of resolution, called tree-like, CDCL (with restarts, deliberately discarding an assignment at hand in order to start from scratch) has been shown to be polynomially equivalent to strictly more powerful general resolution. An excellent source of SAT technology along with its formal foundations is provided by the comprehensive SAT Handbook (Biere et al., 2009). The unique implication point (UIP) traces back to Marques-Silva and Sakallah (1999). The *First-UIP* scheme is discussed in Eén and Sörensson (2004), Marques-Silva et al. (2009), Zhang et al. (2001); its success has been shown both empirically (Dershowitz et al., 2007, Ryan, 2004, Zhang et al., 2001) and analytically (Audemard and Simon, 2009, Pipatsrisawat and Darwiche, 2011).

More unfounded set checking algorithms can be found in Anger et al. (2006), Calimeri et al. (2006), Gebser et al. (2009a), Simons et al. (2002). SAT-based ASP solvers like *assat* (Lin and Zhao, 2004) and *cmodels* (Giunchiglia et al., 2006) add loop formulas to a posteriori eliminate supported models containing some non-empty unfounded set. Stochastic unfounded set checks are investigated in Lin et al. (2006). Unfounded-set-free stable model constructions are explored in Konczak et al. (2006), Linke (2001).

In fact, the rise in complexity in the disjunctive setting reported in Section 2.4.3 is due to more complex unfounded set checking. More precisely, Leone et al. (1997) show that deciding whether a model is free of unfounded sets is *co-NP*-complete. Accordingly, solvers for disjunctive ASP solving rely on two interacting solving components, one for generating stable model candidates and another for unfounded set checking. Disjunctive ASP solving was pioneered by Inoue et al. (1992) and greatly advanced by Leone et al. (2006). Alternative approaches are presented in Giunchiglia et al. (2006), Janhunen et al. (2006). The extension of the above CDNL-based algorithms to a disjunctive setting are described in Drescher et al. (2008).

Algorithm 3 solves the decision problem of stable model existence. Unlike DPLL-style procedures, the transition from computing a single solution to computing multiple or all is non-trivial for

CDCL-based approaches. Although an iterative approach denying previous solutions by recording their complement allows for re-using Algorithm 3 for solution enumeration, it is prone to a significant blow-up in space. Instead, the above algorithmic framework has been extended for enumeration in Gebser et al. (2007b) and projective enumeration in Gebser et al. (2009c). Both enumeration algorithms run in polynomial space due to dedicated backtracking schemes abolishing the need of (persistent) solution recording.

CHAPTER 7

Systems

This chapter focuses on system-specific aspects for enlightening the inner working of ASP systems. Following the workflow in ASP solving, we concentrate on the ASP grounder *gringo* and the ASP solver *clasp* in the first two sections. While *gringo* can be regarded as a realization of the (simplistic) grounding algorithms presented in Chapter 4, *clasp* follows the approach to conflict-driven ASP solving described in Chapter 6. Both constitute central components of *Potassco*, the Potsdam Answer Set Solving Collection, bundling various ASP tools developed at the University of Potsdam. Further selected ASP systems from Potassco are detailed in the third section of this chapter.

All systems described below are freely available as open source packages at potassco. sourceforge.net and distributed under GNU general public license.

7.1 GROUNDING WITH *GRINGO*

The ASP grounder *gringo* was initially conceived as an alternative to the ASP grounder *lparse* and thus shares common input and output formats. While the first versions of *gringo* also imposed input restrictions, guaranteeing finite ground instantiations, only safety is required from version 3 upward.

7.1.1 ARCHITECTURE

Let us begin by describing the high-level architecture of *gringo* by following its workflow in Figure 7.1. The parsing of the input language is done by means of lexer and parser generators. For

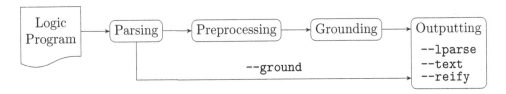

Figure 7.1: Architecture of ASP grounder *gringo*.

processing already ground programs, option --ground allows for bypassing the actual grounding process. Otherwise, preprocessing starts by removing syntactic sugar, like interval .. or pooling ; operators, and deals with grounder directives, like #const or #domain. Moreover, programs are normalized. Among others things, nested terms in positive body literals and positive conditions are factored out. For instance, p(X+X) is replaced with 'p(I), I := X+X'. This process can be understood by comparing Listings 1.3 and 1.4 on Pages 5 and 6 and Listing 7.2 below. Afterward the program is grounded according to Chapter 4 and output in the selected format. While --text provides human readable output, Option --reify turns the program into facts, as detailed in Section 4.3. The default option --lparse results in a machine-oriented format, described below.

7.1.2 *GRINGO'S* INPUT LANGUAGE

The input language of *gringo* follows the syntax of the language constructs given in Section 2.3. This includes normal, choice, cardinality, weight, and disjunctive rules, integrity, cardinality and weight constraints, as well as optimization statements. Also, Section 2.3 contains examples illustrating conditional literals and classical negation.

Hence, we concentrate in what follows on some particular features of *gringo*'s input language complementing the previous chapters, and leave a comprehensive treatment to the user's manual.

Aggregates An *aggregate* is a function on multisets of weights. In combination with comparisons, we obtain aggregate atoms, whose truth values are determined by evaluating the enclosed aggregate. The general form of an *aggregate atom* is as follows.

$$l \ \#A \ [\ \ell_1 = w_1 \ , \ \ldots \ , \ \ell_n = w_n \ ] \ u$$

As with weight constraints in (2.11), all ℓ_i are literals associated with weights w_i that can be specified via arithmetic functions. #A is the name of some function that is applied to the weights of holding literals. Also, l and u are arithmetic functions (or even variables), determining a lower and an upper bound for the result of applying #A. Finally, an aggregate atom is true if the result of applying #A to the weights of holding literals is between l and u (both inclusive). Currently, *gringo* supports the aggregate functions #sum (the sum of weights), #min (the minimum weight), #max (the maximum weight), and #avg (the average of weights).

In addition, there are three more aggregates whose atoms obey a slightly different syntax:

$$l \ \#count \ \{ \ \ell_1 \ , \ \ldots \ , \ \ell_n \ \} \ u$$
$$\#even \ \{ \ \ell_1 \ , \ \ldots \ , \ \ell_n \ \}$$
$$\#odd \ \{ \ \ell_1 \ , \ \ldots \ , \ \ell_n \ \}$$

The #count aggregate is similar to #sum yet with each literal's weight being 1. However, as indicated by curly brackets, the literals of #count are understood as a set, so that duplicates are not counted twice.

For instance, the following aggregate atoms express the same weight constraint.

```
1 #sum    [a=1, not b=1]        1
1 #count {a,    not b}          1
1 #count {a,a, not b,not b} 1
```

Likewise, the literals of parity aggregates #even and #odd, whose aggregate atoms hold if the number of holding literals is even or odd, respectively, are understood as a set.

There are further shorthands that can be used when writing aggregates. With #sum, #min, #max, and #avg, weights of literals can be omitted, in which case 1 is used by default, that is, $\ell_i = 1$

is equivalent to just ℓ_i. Furthermore, either or both of the bounds l and u (applicable to all aggregate atoms but parity) can be omitted, in which case they are taken to be (trivially) satisfied, regardless of an aggregate's evaluation. Finally, the aggregates #sum and #count are used by default for multisets of (weighted) literals in square brackets and sets of literals in curly brackets, respectively. For instance, the following aggregate atoms are synonyms.

```
2 #sum [a, not b, c=2] 3
2       [a, not b, c=2] 3
```

Likewise, we can optionally omit #count in front of curly brackets:

```
1 #count {a, not b} 1
1         {a, not b} 1
```

By tolerating the omission of #sum and #count, weight and cardinality constraints can be written in their traditional notations (without keywords).

For the multiset-based aggregates #sum, #min, #max, and #avg, it is worth noting that they behave differently in case of multiple occurrences of the same literal. For instance, while '2 #sum [a=2] 2' and '2 #sum [a,a] 2' are synonyms, the following pairs of aggregate atoms are different from one another.

```
2 #min [a=2] 2 and 2 #min [a,a] 2
2 #max [a=2] 2 and 2 #max [a,a] 2
2 #avg [a=2] 2 and 2 #avg [a,a] 2
```

That is, whether repetitions of a literal are cumulative or redundant depends on the aggregate being used.

Having discussed aggregate atoms, let us note that there is a second way to use aggregates: the values obtained by evaluating them can be assigned to variables. To this end, *gringo* allows for the aggregates #sum, #min, #max, and #count to occur on the right-hand side of assignment predicate := (or =). For instance, the following rules assign the obtained values to a variable.

```
sum(X) :- X := #sum [a=2,a=3].
min(X) :- X := #min [a=2,a=3].
max(X) :- X := #max [a=2,a=3].
cnt(X) :- X := #count {a,a}.
```

Assuming that atom a holds, we derive sum(5), min(2), max(3), and cnt(1) from the above rules. Otherwise, we get sum(0), min(#supremum), max(#infimum), and cnt(0). Here, the special constants #supremum and #infimum, obtained by applying #min and #max to the empty set of weights, stand for $+\infty$ and $-\infty$, respectively.

Although it seems convenient to use assignments of aggregates' values, this feature should be used with care. If the (weighted) literals of an aggregate belong to domain predicates or built-ins, *gringo* evaluates the aggregate during grounding, thereby obtaining a unique value to be assigned. Otherwise, if the literals do not belong to domain predicates, the value of an aggregate is not known during grounding, in which case *gringo* unwraps all possible outcomes of the aggregate's evaluation.

The latter can lead to a significant blow-up in space, and encoding alternatives without aggregate value assignments may be advantageous.

Having considered ground aggregate atoms, we refer the interested reader to *gringo*'s user manual for details on non-ground aggregates.

Directives   A directive is a meta statement instructing the grounder how to deal with certain parts of the program.

The simplest such directives handle comments. A comment until the end of a line is initiated by symbol %, and a comment within one or more lines is enclosed in %* and *%.

Other directives are preceded by the symbol #.

The "displaying" part of a program allows for suppressing atoms in the output by means of #hide and #show directives. The meaning of the following statements are indicated via accompanying comments.

```
#hide.                % Suppress all atoms in output
#hide p/2.            % Suppress all atoms of predicate p/2 in output
#hide p(X,Y) : q(X).  % Suppress instances of p/2 satisfying condition
```

For selectively including atoms of a certain predicate in the output, one may use the #show directive. Here are some examples:

```
#show p/2.            % Include all atoms of predicate p/2 in output
#show p(X,Y) : q(X).  % Include instances of p/2 satisfying condition
```

A typical usage of #hide and #show is to hide all predicates via #hide and to selectively re-add atoms of certain predicates p/n to the output via '#show p/n'.

Constants appearing in a logic program may serve as placeholders for concrete values to be provided by a user. The #const directive allows us to define a default value to be inserted for a constant. Such a default value can still be overridden via command line option --const. Syntactically, #const must be followed by an assignment having a (symbolic) constant on the left-hand side and a term on the right-hand side. For instance, the declarations '#const x = 42' and '#const y = f(x,h)' make *gringo* turn the fact p(x,y) into p(42,f(42,h)).

#external directives are used to mark certain atoms as external input to the program. This means that such atoms are exempt from simplification and are hence not removed from a program. Typical use cases include the separate grounding of program modules (e.g., with *asptools*) or the dynamic treatment of data streams (e.g., with *oclingo*).

There are two kinds of external directives, global and local external statements.

Global external statements have the form '#external p/n' and mark complete predicates irrespective of any arguments as external. This means that nothing is known about the predicate and hence it cannot be used for instantiation. Consider the following example.

```
#external q/1.
p(1). p(2).
r(X) :- q(X), p(X).
```

The external predicate q/1 is not used for simplification of the problem, and so two ground rules (excluding facts) are obtained.

Local external statements have the form '#external p'. In contrast to global external directives, local ones specify which atoms are external and hence can be used for instantiation. Again, consider a similar example.

```
#external q(X) : p(X).
p(1). p(2).
r(X) :- q(X).
```

The external predicate q/1 is used to bind variable X, yielding the same rules as in the example above.

Integrated scripting language In ASP solving, the grounder is in charge of deterministic computations on the input program. As such, it can be viewed as an extended deductive database system comprising the expressive power of a Turing machine. Nonetheless, certain computations are hard to express with a deductive machinery, and we have to resort to extra-logical means.

For this purpose, the scripting language *lua* is embedded in *gringo*. For example, the greatest common divisor given by instances of predicate p/1 can be calculated via *lua* and captured in the third argument of predicate q/3 as follows.

```
#begin_lua
  function gcd(a,b)
    if a == 0 then return b else return gcd(b%a,a) end
  end
#end_lua.

p(2*3*5;2*3*7;2*5*7).

q(X,Y,gcd(X,Y)) :- p(X;Y), X < Y.
```

When passing this program to *gringo*, it calculates the numbers being arguments of predicate p/1, viz. 30, 42, and 70, while the implementation of the gcd function in *lua* is used to derive q(30,42,6), q(30,70,10), and q(42,70,14).

Beyond sophisticated arithmetic, *lua* also allows for environment interaction. For instance, it provides interfaces to read values from a database. In the following example, we use *sqlite3*, as it is embedded in precompiled *gringo* binaries.

```
1  #begin_lua
2    local env  = luasql.sqlite3()
3    local conn = env:connect("db.sqlite3")
4    function query()
5      local cur = conn:execute("SELECT * FROM test")
6      local res = {}
7      while true do
8        local row = {}
9        row = cur:fetch(row,"n")
10       if row == nil then break end
11       res[#res + 1] = Val.new(Val.FUNC,row)
12     end
13     cur:close()
14     return res
15   end
16 #end_lua.

18 p(X,Y) :- (X,Y) := @query().
```

We define a *lua* function query to read data from an external database table test. Although we do not delve into details here, we draw the reader's attention to Line 11. If test contains tuples ⟨1,a⟩, ⟨2,b⟩, and ⟨3,c⟩, they are successively inserted into the array res. The collected tuples are then taken to construct the facts p("1","a"), p("2","b"), and p("3","c").

Moreover, *lua* can be used within *clingo* for intercepting stable models or interacting between grounding and solving when proceeding incrementally with *iclingo*.

7.1.3 MAKING GROUNDING MORE TRANSPARENT

gringo offers several features for making its grounding process more transparent. The following examples use *gringo* version 3.

The first option of interest is --gstats providing statistics about the internal representation of the input program. To illustrate this, we give in Listing 7.1 the statistics obtained when grounding the Towers of Hanoi encoding in Listing 1.4 along with the instance in Listing 1.3 (cf. Page 6 and 5).

Listing 7.1: Grounding Towers of Hanoi with extended grounder statistics (--gstats)

```
$ gringo tohI.lp tohE.lp --gstats > /dev/null
=== Grounder Input Statistics ===
components              :    20
 non-trivial            :     0

predicates              :     8
 visible                :     1
 average parameters     :   2.1
```

```
statements              :    17
  rules                 :    10
  facts                 :     3
  constraints           :     4
  optimize              :     0

body literals           :    31
  literals in aggregates :    3
  relations             :     6
  positive predicates   :    17
  negative predicates   :     2

aggregates              :     2
  count                 :     2
  sum                   :     0
  avg                   :     0
  even/odd              :     0
  min/max               :     0

terms                   :    96
  variable terms        :    69
  restricted terms      :     7
  constant terms        :    20
```

Note that the statistics are output via the standard error stream in order to avoid interfering with the actual grounding result that is sent to the standard output stream (and here hidden via a redirection to /dev/null).

The internal representation of the encoding in Listing 1.4 (together with the instance in Listing 1.3) is given in Listing 7.2.

Listing 7.2: Internal program representation of the Towers of Hanoi encoding in Listing 1.4

```
$ gringo tohI.lp tohE.lp --verbose > /dev/null
% disk(#I0):-#range(#I0,1,4).
% init_on(#I0,a):-#range(#I0,1,4).
% goal_on(#I0,c):-#range(#I0,1,4).
% 1 #count{move(D,P,T):peg(P):disk(D)} 1:-T:=#I0,#range(#I0,1,15).
% move(D,T):-move(D,_,T).
% on(D,P,0):-init_on(D,P).
% on(D,P,T):-move(D,P,T).
% on(D,P,T+1):-on(D,P,T),not move(D,T+1),T<15.
% blocked(D-1,P,T+1):-on(D,P,T),T<15.
% blocked(D-1,P,T):-blocked(D,P,T),disk(D).
% :-move(D,P,T),blocked(#I0,P,T),#I0:=D-1.
% :-move(D,T),on(D,P,#I0),blocked(D,P,T),#I0:=T-1.
% :-not 1 #count{on(D,P,T)} 1,disk(D),T:=#I0,#range(#I0,1,15).
% :-goal_on(D,P),not on(D,P,15).
```

The major changes deal with the expansion of constant moves by 15 and of range definitions via the internal predicate #range/3. For example, the resulting program has ten (non-factual) rules, three facts, four (integrity) constraints along with two count aggregates, as detailed in Listing 7.1. Also, we get 96 occurrences of terms among which we find 69 occurrences of variables.

Although all rules are prefixed with the comment character %, they are also written to standard error for a clear separation from the grounding result. In fact, grounding with option --verbose allows for getting a visual impression on how long it takes to ground individual rules. This is illustrated in the abridged Listing 7.3.

Listing 7.3: Verbosely grounding Towers of Hanoi (--verbose)

```
$ gringo tohI.lp tohE.lp --verbose --text

[...]

% disk(#I0):-#range(#I0,1,4).
disk(1).
disk(2).
disk(3).
disk(4).

[...]

% 1 #count{move(D,P,T):peg(P):disk(D)} 1:-T:=#I0,#range(#I0,1,15).
1 #count{move(4,c, 1),move(3,c, 1),  [...]  ,move(2,a, 1),move(1,a, 1)} 1.
1 #count{move(4,c, 2),move(3,c, 2),  [...]  ,move(2,a, 2),move(1,a, 2)} 1.
1 #count{move(4,c, 3),move(3,c, 3),  [...]  ,move(2,a, 3),move(1,a, 3)} 1.
1 #count{move(4,c, 4),move(3,c, 4),  [...]  ,move(2,a, 4),move(1,a, 4)} 1.
1 #count{move(4,c, 5),move(3,c, 5),  [...]  ,move(2,a, 5),move(1,a, 5)} 1.
1 #count{move(4,c, 6),move(3,c, 6),  [...]  ,move(2,a, 6),move(1,a, 6)} 1.
1 #count{move(4,c, 7),move(3,c, 7),  [...]  ,move(2,a, 7),move(1,a, 7)} 1.
1 #count{move(4,c, 8),move(3,c, 8),  [...]  ,move(2,a, 8),move(1,a, 8)} 1.
1 #count{move(4,c, 9),move(3,c, 9),  [...]  ,move(2,a, 9),move(1,a, 9)} 1.
1 #count{move(4,c,10),move(3,c,10),  [...]  ,move(2,a,10),move(1,a,10)} 1.
1 #count{move(4,c,11),move(3,c,11),  [...]  ,move(2,a,11),move(1,a,11)} 1.
1 #count{move(4,c,12),move(3,c,12),  [...]  ,move(2,a,12),move(1,a,12)} 1.
1 #count{move(4,c,13),move(3,c,13),  [...]  ,move(2,a,13),move(1,a,13)} 1.
1 #count{move(4,c,14),move(3,c,14),  [...]  ,move(2,a,14),move(1,a,14)} 1.
1 #count{move(4,c,15),move(3,c,15),  [...]  ,move(2,a,15),move(1,a,15)} 1.

[...]
```

Another static view on the input program (as given in Listing 7.2) is obtained by generating the underlying predicate-rule dependency graph. (The format follows the one described in Section 4.1.) This can be done by invoking the command gringo tohI.lp tohE.lp --dep-graph=tohDG.dot. This command produces the file tohDG.dot containing the graph in Figure 7.2 in the format of the graph description language DOT.

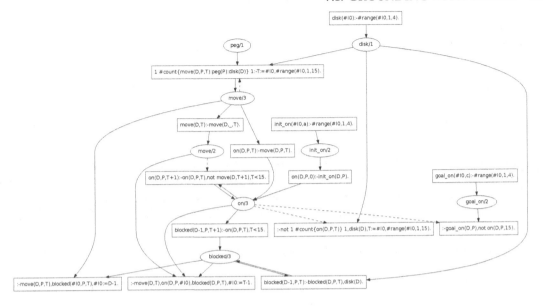

Figure 7.2: Predicate-rule dependency graph of the Towers of Hanoi encoding in Listing 1.4.

7.1.4 THE *SMODELS* FORMAT

The *smodels* format provides an intermediate format for passing a ground logic program from an ASP grounder to an ASP solver. To illustrate this, let us reconsider in Listing 7.4 the human readable version of the ground form of Program `easy.lp` (copied from Listing 4.5).

Listing 7.4: The ground program in Listing 4.5 (on Page 62)

```
1   #count{p(1),p(2),p(3)}.
2   q(2):-p(2),p(3).
3   p(3):-q(2).
4   :-#count{p(3),p(2),p(1)}2.
```

The machine readable version of Listing 7.4 is given in Listing 7.5. By default, *gringo* outputs the resulting ground program in *smodels* format (or via `--lparse`).

Listing 7.5: Internal output of grounding Program `easy.lp` (in Listing 4.4 on Page 62)

```
0   $ gringo easy.lp
1   3 3 2 3 4 0 0
2   1 5 2 0 3 4
3   1 4 1 0 5
4   2 6 3 0 3 4 3 2
5   1 1 1 1 6
6   0
7   2 p(1)
```

```
 8 || 3  p(2)
 9 || 4  p(3)
10 || 5  q(2)
11 || 0
12 || B+
13 || 0
14 || B-
15 || 1
16 || 0
17 || 1
```

In this format, each ground atom is identified through a unique positive integer. Number 0 is used as a separator. Number 1 is an internal atom standing for _false (and preassigned to F, as detailed below). Visible atoms are mapped to their symbolic representation via a symbol table (cf. Lines 7–10). The visibility of atoms occurring in a program is controlled by #hide and #show directives. In what follows, we use ι to represent the mapping of atoms to numbers, regardless of whether they are visible or not. In our example, we have $\iota(\_false) = 1$, $\iota(p(1)) = 2$, $\iota(p(2)) = 3$, $\iota(p(3)) = 4$, and $\iota(q(2)) = 5$.

The first part of a ground program in *smodels* format consists of its rules followed by 0 (see Lines 1–6). Each line starts with a number identifying the type of rule, according to the schema in Table 7.1. As indicated by the leading 3, the sequence '3 3 2 3 4 0 0' in Line 1 accounts for

Table 7.1: The *smodels* format
Type/Format
Normal rule (2.1), Page 13
$1 \lrcorner \iota(a_0) \lrcorner n \lrcorner n - m \lrcorner \iota(a_{m+1}) \lrcorner \ldots \lrcorner \iota(a_n) \lrcorner \iota(a_1) \lrcorner \ldots \lrcorner \iota(a_m)$
Cardinality rule (2.4), Page 18
$2 \lrcorner \iota(a_0) \lrcorner n \lrcorner n - m \lrcorner l \lrcorner \iota(a_{m+1}) \lrcorner \ldots \lrcorner \iota(a_n) \lrcorner \iota(a_1) \lrcorner \ldots \lrcorner \iota(a_m)$
Choice rule (2.3), Page 18
$3 \lrcorner m \lrcorner \iota(a_1) \lrcorner \ldots \lrcorner \iota(a_m) \lrcorner 0 - m \lrcorner 0 - n \lrcorner \iota(a_{n+1}) \lrcorner \ldots \lrcorner \iota(a_o) \lrcorner \iota(a_{m+1}) \lrcorner \ldots \lrcorner \iota(a_n)$
Weight rule (2.10), Page 21
$5 \lrcorner \iota(a_0) \lrcorner l \lrcorner n \lrcorner n - m \lrcorner \iota(a_{m+1}) \lrcorner \ldots \lrcorner \iota(a_n) \lrcorner \iota(a_1) \lrcorner \ldots \lrcorner \iota(a_m) \lrcorner w_{m+1} \lrcorner \ldots \lrcorner w_n \lrcorner w_1 \lrcorner \ldots \lrcorner w_m$
Minimize statement[1] (2.12), Page 22
$6 \lrcorner 0 \lrcorner n \lrcorner n - m \lrcorner \iota(a_{m+1}) \lrcorner \ldots \lrcorner \iota(a_n) \lrcorner \iota(a_1) \lrcorner \ldots \lrcorner \iota(a_m) \lrcorner w_{m+1} \lrcorner \ldots \lrcorner w_n \lrcorner w_1 \lrcorner \ldots \lrcorner w_m$
Disjunctive rule (2.14), Page 24
$8 \lrcorner m \lrcorner \iota(a_1) \lrcorner \ldots \lrcorner \iota(a_m) \lrcorner 0 - m \lrcorner 0 - n \lrcorner \iota(a_{n+1}) \lrcorner \ldots \lrcorner \iota(a_o) \lrcorner \iota(a_{m+1}) \lrcorner \ldots \lrcorner \iota(a_n)$

the choice rule obtained from #count{p(1),p(2),p(3)} (following the transformation in (2.8) on Page 20). The second number, 3, tells us that we have the choice among three atoms, whose identifiers, 2, 3, 4, follow in the sequence. The two trailing 0s indicate that there are no body literals. As designated by the first number, 1, the sequence '1 5 2 0 3 4' stands for a normal rule, viz.

---

[1]Assuming that $\ell_i = a_i$ for $1 \leq i \leq m$ and $\ell_i = \sim a_i$ for $m + 1 \leq i \leq n$.

'q(2) :- p(2),p(3)'. The second number gives the head identifier, $\iota(q(2)) = 5$, followed by the number of body literals, 2, among which we find 0 negative ones. The last two numbers give the identifiers of the (positive) body literals, viz. $\iota(p(2)) = 3$ and $\iota(p(3)) = 4$. Analogously, '1 4 1 0 5' stands for 'p(3) :- q(2)'.

To capture the last rule in Listing 7.4, we must first convert it into admissible rule types in *smodels* format. To this end, we use transformation (2.6) (on Page 20) for turning ':- #count{p(3),p(2),p(1)} 2' into a cardinality rule along with an integrity constraint. Let us indicate the intermediate result by the following pseudo code using the auxiliary variable _aux with $\iota(\_aux) = 6$.

```
4   _aux :- 3 #count{p(3),p(2),p(1)}.
5   _false :- not _aux.
```

The two previous rules are captured by Lines 4 and 5 of Listing 7.5. To be more precise, the initial number 2 tells us that '2 6 3 0 3 4 3 2' captures a cardinality rule, having head identifier 6, 3 contained literals, among which 0 are negative, lower bound 3, and body literals having identifiers 4, 3, and 2. Although the last rule is a byproduct of the aforementioned transformation, it is instructive to realize that it reflects the encryption of an integrity constraint. At first glance, '1 1 1 1 6' captures a normal rule, having $\iota(\_false) = 1$ as head identifier along with a single negative body literal, identified by $\iota(\_aux) = 6$. However, the special-purpose atom _false is always preassigned to F, as can be read off the third part of Listing 7.5.

Actually the symbol table in Lines 7–10 (terminated by 0 in Line 11) in Listing 7.5 is followed by so-called compute statements, preassigning truth values to atoms. Atom identifiers between B+ and 0 are assigned T, whereas the ones between B− and 0 are assigned F. In our case, only atom identifier 1 is set to F, attributing all rules with head atom _false the character of an integrity constraint. The final entry in Listing 7.5, viz. 1 in Line 17, indicates how many models should be computed.

7.1.5 OUTLOOK

Our account of *gringo* including all examples deals with version 3.0.4. Currently, a new version of *gringo* is implemented, which will lead to *gringo* series 4. Apart from an entire re-implementation, this construction series will also feature a significantly extended and revised input language, among other things, comprising the proposal for the ASP input language *ASP-Core-2*.

7.2 SOLVING WITH *CLASP*

The solver *clasp* is originally designed and optimized for conflict-driven ASP solving. To this end, it features a number of sophisticated reasoning and implementation techniques, some specific to ASP and others borrowed from CDCL-based SAT solvers. While *clasp* can also be used as a full-fledged SAT, MaxSAT[2], or PB[3] solver, we illustrate its major features from the perspective of ASP solving.

Although we detail *clasp*'s architecture further below, it might be instructive to accompany further reading by regarding Figure 7.3 on Page 129.

7.2.1   INTERFACES AND PREPROCESSING

For ASP solving, *clasp* reads propositional logic programs in *smodels* format.[4] [5] Choice rules as well as cardinality and weight constraints are either compiled into normal rules during parsing, configured via the option `--trans-ext`, or (by default) dealt with in an intrinsic fashion. Although we do not detail this here, we note that the intrinsic treatment of cardinality and weight constraints augments the source-pointer-based unfounded set algorithm in Section 6.5 and requires care during program simplifications.

At the beginning, a logic program is subject to extensive preprocessing. The idea is to simplify the program while identifying equivalences among its relevant constituents. These equivalences are then used for building a compact program representation (in terms of Boolean constraints). Logic program preprocessing is configured via option `--eq`, taking an integer value fixing the number of iterations. (Interestingly, preprocessing is sometimes able to turn a non-tight program into a tight one.) Once a program has been transformed into Boolean constraints, the nogoods among them are (optionally) subject to further, mostly SAT-based preprocessing. This type of preprocessing is invoked with option `--sat-prepro` and further parameters. Such techniques are however more involved in ASP because variables relevant to unfounded set checking, optimization, or part of cardinality and weight constraints cannot be simply eliminated. Note that both preprocessing steps identify redundant variables that can be expressed in terms of the relevant ones included in the resulting set of constraints.

A major yet internal feature of *clasp* is that it can be used in a stateful way. That is, *clasp* may keep its state, involving program representation, recorded nogoods, heuristic values, etc., and be invoked under additional (temporary) assumptions and/or by adding new atoms and rules. The corresponding interfaces are fundamental for supporting incremental and reactive ASP solving as realized in *iclingo* and *oclingo*, respectively. Furthermore, solving under assumptions is essential to parallel solving, as done in *clasp*'s multi-threaded distributed search as well as its cluster-oriented variant *claspar*.

---

[2]MaxSAT stands for the Maximum SAT problem.
[3]PB stands for Pseudo-Boolean.
[4]Currently, disjunctive logic programs are only handled by *claspd* (see Section 7.3.1).
[5]Further input formats accepted by *clasp* include (extended) *dimacs*, *opb*, and *wbo*.

7.2.2 REASONING MODES

Although *clasp*'s primary use case is the computation of stable models, it also allows for computing supported models of logic programs via option `--supp-models`.[6] In addition, *clasp* provides a number of reasoning modes, determining how to proceed when a model is found.

Enumeration Solution enumeration is non-trivial in the context of backjumping and conflict-driven learning. A simple approach relies on recording solutions as nogoods and exempting them from deletion. Although *clasp* supports this via option `--enum-mode=record`, it is prone to blow-up in space in view of a possible exponential number of solutions. In contrast, the default enumeration algorithm of *clasp* runs in polynomial space. Both enumeration approaches also allow for projecting models to a subset of atoms, invoked with `--project` and configured via directives `#hide` and `#show`. This option is of great practical value whenever one faces overwhelmingly many models, involving solution-irrelevant variables having proper combinatorics. For example, the program consisting of the choice rule `{a,b,c}` has eight stable models. When augmented with directive '`#hide c`', still eight solutions are obtained, yet including four duplicates. In contrast, invoking *clasp* with `--project` yields only four stable models differing on a and/or b.

As detailed in Section 7.2.4, *clasp* offers a dedicated interface for enumeration. This allows for abstracting from how to proceed once a model is found and thus makes the search algorithm independent of the concrete enumeration strategy. Further reasoning modes implemented via the enumeration interface admit computing the intersection or union of all stable models of a program (by setting `--enum-mode` to `cautious` or `brave`, respectively). Rather than computing the whole collection of (possibly) exponentially many stable models, the idea is to compute a first stable model, record a constraint eliminating it from further solutions, then compute a second stable model, strengthen the constraint to represent the intersection (or union) of the first two stable models, and to continue like this until no more stable models are obtained. This process involves computing at most as many stable models as there are atoms in an input program. Either the cautious or the brave consequences are then given by the atoms captured by the final constraint.

Optimization An objective function is specified via `#minimize` and/or `#maximize` statements. *clasp* offers several options for finding optimal solutions. First, the objective function can be initialized via `--opt-value`. Second, *clasp* allows for computing one or all (via `--opt-all`) optimal solutions. Such options are useful when one is interested in computing consequences belonging to all optimal solutions (in combination with `--enum-mode=cautious`). To this end, one starts with searching for an (arbitrary) optimal stable model and then re-launches *clasp* by bounding its search with the obtained optimum. Doing the latter with `cautious` yields the atoms that belong to all optimal stable models. Option `--restart-on-model`, making *clasp* restart after each (putatively optimal) solution, turned out to be effective for ameliorating convergence to an optimum. Particular strategies for lexicographic optimization serve the same purpose, especially on large and under-constrained

---

[6]To be more precise, option `--supp-models` disables unfounded set checking. Sometimes the grounder or preprocessing may already eliminate some supported models such that they cannot be recovered later on.

multi-criteria optimization problems. Moreover, option `--opt-heuristic` can be used to alter sign selection (see below) toward a better objective function value. Optimization is also implemented via the aforementioned enumeration interface. When a solution is found, an optimization constraint is updated with the corresponding objective function value.[7] Then, the decision level violating the constraint is identified and retracted, or if the constraint is violated at decision level 0, search terminates. It is also worth mentioning that *clasp* propagates optimization constraints, that is, they can imply (and provide reasons for) literals upon unit propagation. Finally, when optimization is actually undesired and all solutions ought to be inspected instead, option `--opt-ignore` is available to avoid modifying the input (by ignoring optimization statements).

An innovative feature of *clasp* 2 is hierarchical optimization (`--opt-hierarch`), build on top of uniform optimization. Hierarchical optimization allows for solving multi-criteria optimization problems by considering criteria according to their respective priorities. Such an approach is more involved than standard branch-and-bound-based optimization because it must recover from unsatisfiable subproblems, one for each criterion. This is accomplished by dynamic optimization constraints that may be disabled and reinitialized during search. Accordingly, learned nogoods relying on such constraints must be retracted once the constraints get disabled. Another benefit of such dynamic constraints is that they allow for decreasing an (upper) bound in a non-uniform way, and successively re-increasing it upon unsatisfiability.

7.2.3   PROPAGATION AND SEARCH

Propagation in *clasp* relies on a general interface called (Boolean) *Constraint* and is thus not limited to nogoods. However, dedicated data structures are used for binary and ternary nogoods, accounting for the many short nogoods stemming from completion. More complex constraints are accessed via watch lists for each literal, storing Boolean constraints that need to be updated when the variable becomes true or false, respectively. While unit propagation of long nogoods is based on the two-watched literal technique, a counter-based approach is used for propagating cardinality and weight constraints. A literal implied by a Boolean constraint upon unit propagation stores a reference to that constraint, which in turn can be queried for an antecedent.

Motivated by the nature of ASP problems, *clasp* first applies unit propagation to binary and ternary nogoods, then longer nogoods, and finally other constraints. Moreover, as detailed in Algorithm 4 (on Page 101), its propagation procedure gives a clear preference to unit propagation over unfounded set computations. To this end, *clasp* abstracts from its various propagation mechanisms by using post propagators. That is, it maintains a priority list of post propagators that are consecutively processed after unit propagation. For instance, failed-literal detection, unfounded set checking, and parallel search are implemented as post propagators. Similarly, they are used in *clasp*'s extension with constraint processing, *clingcon*, to realize theory propagation.

Unfounded set detection follows Algorithm 5 and aims at small, rather than greatest, unfounded sets. An intrinsic treatment of cardinality and weight constraints augments unfounded set

---

[7]At the implementation level, all optimization constraints are minimize constraints.

detection by means of source pointers, still aiming at lazy unfounded set checking. The representation of loop nogoods is controlled via option `--loops`. With value `distinct`, loop nogoods are generated for individual unfounded atoms, as shown in Algorithm 4. However, the default value `common` makes *clasp* only compute one reason per unfounded set. Like nogoods derived from conflicts, they are subject to unit propagation and deletion (see below). On the other hand, when `--loops=no` is specified, loop nogoods are stored only as long as they serve as antecedents of falsified unfounded atoms.

Decision heuristics   The primary decision heuristics of *clasp* use look-back strategies derived from corresponding approaches in SAT, viz., *berkmin*, *vsids*, and *vmtf*. Such heuristics privilege variables involved in recent conflicts. To this end, they maintain an activity score for each variable, increased upon conflict resolution and decayed periodically. The major difference between the approaches of *berkmin* and *vsids* lies in the scope of variables considered during decision making. While *vsids* selects a free variable that is globally most active, *berkmin* restricts the selection to variables belonging to the most recently recorded yet undispelled dynamic nogood. Although the look-back heuristics implemented in *clasp* are modeled after the corresponding CDCL-based approaches, *clasp* optionally also scores variables contained in loop nogoods. In case of *berkmin*, it may also select a free variable belonging to a recently recorded loop nogood. Finally, we note that *clasp*'s heuristic can also be based upon look-ahead strategies extending unit propagation by failed-literal detection. This makes sense when running *clasp* without conflict-driven nogood learning, operating similar to *smodels*. A limited form of this is available by `--initial-lookahead`.

Once a decision variable has been selected, a sign heuristic decides about its truth value. The main criterion for look-back heuristics is to satisfy the greatest number of conflict nogoods, that is, to pick the literal that occurs in fewer of them.[8] Initially and also for tie-breaking, *clasp* does sign selection based on the type of a variable: atoms are preferably set to false, while bodies are made true. This aims at maximizing the number of resulting implications. Another sign heuristic implemented in *clasp* is progress saving. The idea is to remember truth values of retracted variables upon backjumping (or restarting), except for those assigned at the last decision level. These saved values are then used for sign selection. The intuition behind this strategy is that the literals assigned prior to the last decision level did not lead to a conflict and may have satisfied some subproblem. Hence, re-establishing them may help to avoid solving subproblems multiple times. Progress saving is invoked with option `--save-progress`; its computational impact, however, depends heavily on the structure of a problem at hand (cf. Section 7.2.6).

Restart policies   The robustness of *clasp* is boosted by various restart strategies, namely, arithmetic, geometric, Luby-style, and a nested policy. The first two start with an initial number of conflicts after which *clasp* restarts; this threshold is then increased after each restart, either by addition or multiplication, respectively. For this, `--restarts` takes two arguments, x and y, to make *clasp* follow the restart sequences $x + y \times i$ or $x \times y^i$, respectively, where i counts the number of performed

[8]Satisfaction of nogoods is best understood in terms of the corresponding clauses.

restarts. The Luby-style policy schedules restarts according to a recurrent and progressively growing sequence of numbers of conflicts, viz., $x, x, (2^1 \times x), x, x, (2^1 \times x), (2^2 \times x), x, \ldots$ for some unit x. The nested policy makes restarts follow a two-dimensional pattern that increases geometrically in two dimensions. For this purpose, it takes a third parameter z and repeats the geometric sequence $x \times y^i$ whenever it reaches the outer limit $z \times y^j$, where j counts how often the outer limit was hit so far. Usually, restart strategies are based on the total number of encountered conflicts. Beyond that, *clasp* features local restarts. Here, one counts the number of conflicts per decision level in order to measure the difficulty of subproblems locally. Furthermore, a bounded approach to restarting (and backjumping) is used when enumerating stable models. To complement its more determined search, *clasp* also allows for initial randomized runs, typically with a small restart threshold, in the hope to extract putatively interesting nogoods. Finally, it is worth noting that, despite the fact that recent SAT solvers use rather aggressive restart strategies, *clasp* still defaults to a more conservative geometric policy because it performs better on ASP-specific benchmarks.

Nogood deletion   To limit the number of nogoods stored simultaneously, dynamic nogoods are periodically subject to deletion. Complementing look-back heuristics, *clasp*'s nogood deletion strategy associates an activity with each recorded nogood, which is incremented whenever the nogood is used for conflict resolution. The initial threshold on the number of stored nogoods is calculated from the size of an input program and increased by a certain factor upon each restart. As soon as the current threshold is exceeded, deletion is initiated and removes (by default) up to 75% of the recorded nogoods. This can be modified via --dfrac. Nogoods that are currently locked (because they serve as antecedents) or whose activities significantly exceed the average activity are exempt from deletion. In addition, --dglue allows for keeping nogoods having a literal block distance smaller than or equal to a specified threshold. The remaining nogoods have their activities decayed in order to account for recency of usage. All in all, *clasp*'s nogood deletion strategy aims at limiting the overall number of stored nogoods, while keeping the relevant and recently recorded ones. This likewise applies to conflict and loop nogoods.

7.2.4   MULTI-THREADED ARCHITECTURE

A major extension of *clasp* series *2* lies in its parallel solving capacities. To this end, *clasp* follows a coarse-grained, task-parallel approach via shared memory multi-threading. Given this, *clasp* allows for parallel solving by search space splitting and/or competing strategies. While the former involves dynamic load balancing in view of highly irregular search spaces, both modes aim at running searches as independently as possible in order to take advantage of enhanced sequential algorithms. In fact, a portfolio of solver configurations cannot only be used for competing but also in splitting-based search. The latter is optionally combined with global restarts to escape from uninformed initial splits.

In what follows, we focus on the multi-threaded component and communication architecture of *clasp*.

Component architecture   To explain the architecture and functioning of *clasp*'s multi-threaded architecture, let us follow the workflow underlying its design. To this end, consider *clasp*'s architectural diagram given in Figure 7.3.

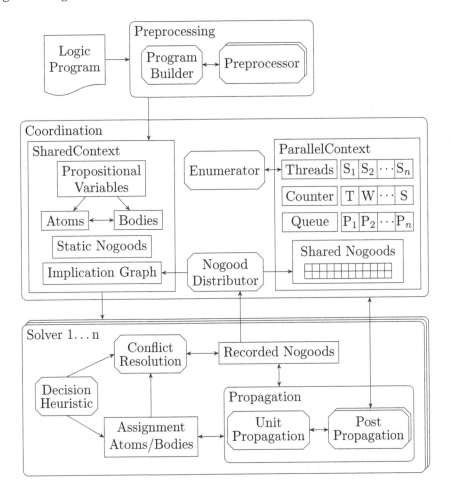

Figure 7.3: Architecture of *clasp* 2.

At the start, only the main thread is active. Once a logic program is read in, all aforementioned preprocessing stages are conducted by this thread. The outcome of the preprocessing phase is stored in a *SharedContext* object that is initialized by the main thread and shared among all participating threads. Among others, this object contains

- the set of relevant Boolean variables together with type information (e.g., atom, body, aggregate, etc.),

- a symbol table, mapping (named) atoms from the program to internal variables,

- the positive atom-body dependency graph, restricted to its (non-trivial) strongly connected components,

- the set of Boolean constraints, among them nogoods, cardinality and weight constraints, optimization constraints, and

- an implication graph capturing inferences from binary and ternary nogoods.[9]

The richness of this information is typical for ASP, and it is much sparser in a SAT setting.

After its initialization in conjunction with the main solver, further (solver) threads are (concurrently) attached to the *SharedContext*, where its constraints are "cloned." Notably, each constraint is aware of how to clone itself efficiently. Moreover, the *Enumerator* and *NogoodDistributor* objects are used globally in order to coordinate various model enumeration modes and nogood exchange among solver instances. We detail their functioning below.

Each thread contains one *Solver* object, implementing an extended version of Algorithm 3. Every *Solver* stores

- local data, including assignment, watch lists, constraint database, etc.,

- local strategies, regarding heuristics, restarts, constraint deletion, etc.,

and it uses the *NogoodDistributor* to share recorded nogoods. As mentioned above, each solver propagates first binary and ternary nogoods (shared through the aforementioned implication graph), then longer nogoods and other constraints, before it finally applies any available post propagators.

ParallelContext controls concurrent solving with (up to 64) individually configurable threads (via option --threads). When attaching a solver to the *SharedContext*, *ParallelContext* associates a thread with the solver and adds dedicated post propagators to it. One high-priority post propagator is added for message handling and another, very low-priority post propagator is supplied for integrating information stemming from models[10] and/or shared nogoods.

To control parallel search, *ParallelContext* maintains a set of atomic message flags:

- *terminate* signals the end of a computation,

- *interrupt* forces outside termination (e.g., when the user hits Ctrl+C),

- *sync* indicates that all threads shall synchronize, and

- *split* is set during splitting-based search whenever at least one thread needs work.

These flags are used to implement *clasp*'s two major parallel search strategies:

---

[9]Note that unary nogoods capture initial problem simplifications that need not be rechecked during search.
[10]This can regard an enumerated model to exclude, intersect, or union, as well as objective function values.

- *splitting-based search* via distribution of guiding paths (that is, the sequence of all non-deterministic choices) and dynamic load balancing via a split-request and -response protocol, and

- *competition-based search* via freely configurable solver portfolios.

Notably, solver portfolios can also be used in splitting-based search, that is, different guiding paths can be solved with different configurations. Their usage is invoked through option `--portfolio`; an example specification is obtained with `--create-template`.

Communication architecture   A salient transverse aspect of the multi-threaded architecture of *clasp* 2 is its communication infrastructure, used for implementing advanced reasoning procedures. To begin with, the *ParallelContext* object keeps track of threads' load, particularly in splitting-based search. Moreover, the *Enumerator* controls enumeration-based reasoning modes, while the *NogoodDistributor* handles the exchange of recorded nogoods among solver threads. These communication-intense components along with some implementation techniques are detailed below in increasing order of complexness.

Thread coordination   The basic communication architecture of *clasp* relies on message passing, efficiently implemented by lock-free atomic integers. On the one hand, globally shared atomic counters are stored in *ParallelContext*. For instance, all aforementioned control flags are stored in a single shared atomic integer. On the other hand, each thread has a local message counter hosted by the message handling post propagator. Message passing builds upon two basic methods: `postMessage()` and `hasMessage()`. Posting a message amounts to a Compare-and-Swap[11] on an atomic integer, and checking for messages (via specialized post propagators) is equivalent to an atomic read. Of particular interest is communication during splitting-based search. This is accomplished via a lock-free work queue, an atomic work request counter, and a work semaphore in *ParallelContext*. Initially, the work queue only contains the empty guiding path, and all threads "race" for this work package by issuing a work request. A work request first tries to pop a guiding path from the work queue and returns upon success. Otherwise, the work request counter is incremented and a split request is posted, which results in raising the *split* flag. Afterward, a `wait()` is tried on the work semaphore. If `wait()` fails because the number of idle threads now equals the total number of threads, the requesting thread posts a *terminate* message and wakes up all waiting threads. Otherwise, the thread is blocked until new work arrives. On the receiver side, the message handling post propagator of each thread checks whether the *split* flag has been set. If so, and provided that the thread at hand has work to split, its message handler proceeds as follows. At first, it decrements the work request counter. (Note that the message handler thus declares the request as handled before actually serving it in order to minimize over-splitting.) If the work request counter reached 0, the message handler also resets the *split* flag. Afterward, the search space is split and a (short) guiding path is pushed

---

[11] Conditional writing is performed as atomic CPU instruction to achieve synchronization in multi-threading.

to the work queue in *ParallelContext*. At last, the message handler signals the work semaphore and hence eventually wakes up a waiting thread.

Splitting-based search usually suffers from uninformed early splits of the search space. To counterbalance this, *ParallelContext* supports an advanced global restart scheme based on a two-phase strategy (configured through option `--global-restarts`). In the first phase, threads vote upon effectuating a global restart based on some given criterion (currently, number of conflicts); however, individual threads may veto a global restart. For instance, this may happen in enumeration when a first model is found during this first restarting phase. Once there are enough votes, a global restart is initiated in the second phase. For this, a *sync* message is posted and threads wait until all solvers have reacted to this message. The last reacting thread decides on how to continue. If no veto was issued, the global restart is executed. That is, threads give up their guiding paths, the work queue is cleared, and the initial (empty) guiding path is again added to the work queue. Otherwise, the restart is abandoned, and the threads simply continue with their current guiding paths.

If splitting-based search is not active (i.e., during competition-based search), the work queue initially contains one (empty) guiding path for each thread, and additional work requests simply result in the posting of a *terminate* message.

Nogood exchange   Given that each thread implements conflict-driven search involving nogood learning, the corresponding solvers may benefit from a controlled exchange of their recorded information. However, such an interchange must be handled with great care because each individual solver may already learn exponentially many nogoods, so that their additional sharing may significantly hamper the overall performance.

To differentiate which nogoods to share, *clasp* 2 pursues a hybrid approach regarding both nogood exchange and storage. As described above, the binary and ternary implication graph (as well as the positive atom-body dependency graph) are shared among all solver threads. Otherwise, each solver maintains its own local nogood database. The sharing of these nogoods is optional, as we detail next.

The actual exchange of nogoods is controlled in *clasp* by separate distribution and integration components for carefully selecting the spread constraints. This is supported by thread-local interfaces along with the global *NogoodDistributor* (see Figure 7.3). All components rely on interfaces abstracting from the specific sharing mechanism used underneath.

The distribution of nogoods is configurable in two ways (by using option `--distribute`). First, the exported nogoods can be filtered by their type, viz. conflict, loop, or short (i.e., binary and ternary), or be exhaustive or inhibited. The difference between globally sharing short nogoods (via the implication graph) and additionally "distributing" them lies in the proactiveness of the process. While the mere sharing leaves it to each solver to discover nogoods added by others, their explicit distribution furthermore communicates this information through the standard distribution process. Second, the export of nogoods is subject to their respective literal block distance. Fewer distinct decision levels are regarded as advantageous since they are prone to prune larger parts of the search space.[12]

The integration of nogoods is likewise configurable in two ways (by using option --integrate). The first criterion captures the relevance of a nogood to the local search process. First, the state of a nogood is assessed by checking whether it is satisfied, violated, open (i.e., neither satisfied nor violated), or unit with respect to the current (partial) assignment. While violated and unit nogoods are always considered relevant, open nogoods are optionally passed through a filter using the solver's current heuristic values to discriminate the relevance of the candidate nogood to the current solving process. Finally, satisfied nogoods are either ignored or considered open depending on the configuration of the corresponding filter and their state relative to the original guiding path. The second integration criterion is expressed by a grace period influencing the size of the local import queue and thereby the minimum time a nogood is retained. Once the local import queue is full, the least recently added nogood is evicted and either transferred to the thread's nogood database (where it becomes subject to the thread's nogood deletion policy) or immediately discarded. Currently, two modes are distinguished: the thread transfers either all or only "heuristically active" nogoods from its import queue while discarding all others.

Both distribution and integration are implemented as dedicated (complex) post propagators, based upon a global distribution scheme implemented via an efficient lock-free Multi-Read-Multi-Write (MRMW) list situated in *ParallelContext*.[13] Distribution roughly works as follows. When the solver of thread i records a nogood that is a candidate for sharing, it is first integrated into the thread-local nogood database. In addition, the nogood's reference counter is set to the total number of threads plus one, and its target mask to all threads except i. At last, thread i appends the shared nogood to the aforementioned MRMW list.

Conversely upon integration, thread j traverses the MRMW list, thereby ignoring all nogoods whose target mask excludes j. Depending on the state of a nogood, the aforementioned filters decide whether a nogood is relevant or not. All relevant nogoods are integrated into the search process of thread j and added to its local import queue. The reference counter of a nogood is decremented by each thread moving its read pointer beyond it. In addition, the sharing thread i decrements a nogood's reference counter when it no longer uses it. Hence, the reference counter of a shared nogood can only drop to zero once it is no longer addressed by any read pointer. This makes it subject to deletion.

Notably, the shared representation of a nogood is only created when the nogood is actually distributed. Otherwise, its optimized (single-threaded) representation is used. Upon integration, the "best" representation is selected, for instance, short nogoods are copied while longer ones are physically shared.

---

[12]This criterion has empirically shown to be rather effective and largely superior to a selection by length.
[13]This choice is motivated by the fact that we aim at optimizing *clasp* for desktop computers, still mostly possessing few genuine processing units. Other strategies are possible and an active subject of current research.

Complex reasoning modes In addition to model printing, all enumeration-based reasoning modes of *clasp* 2 are controlled by the global *Enumerator* (see Figure 7.3). These reasoning modes include regular and projected model enumeration, intersection and union of models, uniform and hierarchical (multi-criteria) optimization, as well as combinations thereof, like computing the intersection of all optimal models.

As already mentioned, one global *Enumerator* is shared among all threads and is protected by a lock. Whenever applicable, it hosts global constraints, like optimization constraints, that are updated whenever a model is found. Additionally, the *Enumerator* adds a local enumeration-specific constraint to each solver for storing thread-local data, e.g., current optima (see below). Once a model is found, a dedicated message *update-model* is send to all threads, but threads only react to the most recent one.

In fact, enumeration is combinable with both splitting and competing search strategies, either by dedicated enumeration algorithms taking advantage of guiding paths or by using solution recording in a competitive setting. The latter setting exploits the infrastructure for nogood exchange in order to distribute solutions among solver threads. Once a solution is converted into a nogood, it can be treated as usual, except that its integration is imperative and that it is exempt from deletion. However, this approach suffers from exponential space complexity in the worst case. Unlike this, splitting-based enumeration runs in polynomial space, following a distributed version of *clasp*'s dedicated enumeration algorithm. In order to avoid uninformed splits at the beginning, all solver threads may optionally start in a competitive setting. Once the first model is found, the *Enumerator* enforces splitting-based search among all solver threads and disables global restarts. In addition to the distribution of disjoint guiding paths, backtrack levels are dealt with locally in order to guarantee an exhaustive and duplicate-free enumeration of all models.

In optimization, solver threads cooperate in enumerating one better model after another until no better one is found, so that the last model is optimal. Whenever a better model is found, its objective value is stored in the *Enumerator*. The threads react upon the following *update-model* message by integrating the new value into their local optimization constraint representation[14] and thus into the search processes of their solvers. Optimization constraints provide methods for efficiently re-computing their state after an update, so that restarting search is unnecessary in most cases.

Also, brave and cautious reasoning, computing the union and intersection of all models, respectively, are implemented through a global constraint within the *Enumerator*.

7.2.5 MAKING SOLVING MORE TRANSPARENT

The command line option ``--stats`` of *clasp* allows for obtaining a multitude of statistics gathered during the solving process. The output in Listing 7.6 extends the basic one shown in Listing 1.5 on Page 8; in both cases *clasp* is used to solve the Towers of Hanoi puzzle encoded in Listing 1.3 and 1.4.

---

[14]While the literals of an optimization constraint are stored globally, corresponding upper bounds are local to threads, and changes are communicated through the *Enumerator*.

Listing 7.6: Solving Towers of Hanoi with extended solver statistics (--stats=2)

```
$ gringo tohI.lp tohE.lp | clasp --stats=2
clasp version 2.0.5
Reading from stdin
Solving...
Answer: 1
move(4,b, 1) move(3,c, 2) move(4,c, 3) move(2,b, 4) move(4,a, 5) \
move(3,b, 6) move(4,b, 7) move(1,c, 8) move(4,c, 9) move(3,a,10) \
move(4,a,11) move(2,c,12) move(4,b,13) move(3,c,14) move(4,c,15)
SATISFIABLE

Models       : 1+
Time         : 0.011s (Solving: 0.00s 1st Model: 0.00s Unsat: 0.00s)
CPU Time     : 0.000s
Choices      : 345
Conflicts    : 134
Restarts     : 1

Atoms        : 837
Rules        : 1484    (1: 1319 2: 150 3: 15)
Bodies       : 1263
Equivalences : 1685    (Atom=Atom: 87 Body=Body: 69 Other: 1529)
Tight        : Yes

Variables    : 1206    (Eliminated:    0 Frozen:  496)
Constraints  : 1844    (Binary: 61.9% Ternary: 27.1% Other: 11.0%)
Lemmas       : 134     (Binary: 12.7% Ternary: 24.6% Other: 62.7%)
  Conflict   : 134     (Average Length: 4.7)
  Loop       : 0       (Average Length: 0.0)
  Other      : 0       (Average Length: 0.0)
  Deleted    : 0

Backtracks          : 0
Backjumps           : 134    (Bounded: 0)
Skippable Levels    : 341    (Skipped: 341 Rate: 100.0%)
Max Jump Length     : 15     (Executed: 15)
Max Bound Length    : 0
Average Jump Length : 2.5    (Executed: 2.5)
Average Bound Length: 0.0
Average Model Length: 1.0
```

The label Time provides overall wall clock time as measured by *clasp*; it starts with reading the input file. Hence, the discrepancy between global wall clock time and Solving time stands for the time needed for reading in the ground program. A large difference may indicate a grounding problem. Of interest is also the relation between the number of Choices and Conflicts. While similar numbers indicate a highly combinatorial problem, many more Choices than Conflicts point to extensive backjumping.

The following group of figures concerns the input program. The number of rules is broken down into rule types in *smodels* format (cf. Table 7.1 on Page 122). That is, the read ground program contained 1319 normal, 150 cardinality, and 15 choice rules. The label Equivalences gives the number of equivalences between atoms, bodies, and both of them detected during preprocessing.

The next group of indicators describes the Boolean constraints obtained from the program as well as learned nogoods. In our example, the original 837 atoms and 1263 bodies are captured by 1206 Boolean variables. Among them, 496 are excluded from SAT-style preprocessing due to their specific role in ASP solving. Similarly, the 1484 rules of the input program induce 1844 Boolean constraints. Their majority consists of nogoods obtained from the program's completion; among them, more than 1100 binary nogoods (cf. (5.10) and (5.11) on Page 83). Further Boolean constraints result from the dedicated treatment of aggregates. The label Lemmas summarizes the distribution of learned nogoods. Given that our program is tight, no Loop nogoods are generated. Similarly, *clasp* was run in single threaded mode, so that no nogoods were contributed by Other threads. Thus, all 134 learned nogoods were obtained from Conflicts and never Deleted in our simple case.

The last collection of figures reflects the search process. The format is meant to capture not only the search for one stable model but moreover their enumeration. In our example, we observe 134 backjumps with a maximum of 15 and an average of 2.5 decision levels.

The statistics in Listing 7.6 summarize key features of the solving process. For obtaining more structural information, we may appeal to visualization techniques providing insights into the internal structure of a problem. As an example, we give in Figure 7.4 the (reduced) atom interaction graph of the ground program obtained from the Towers of Hanoi problem encoded in Listing 1.3 and 1.4. The graph's vertices consist of all atoms unassigned after preprocessing; two atoms are linked whenever they jointly occur in some nogood. The color gives the accumulated heuristic score of each variable. That is, red reflects a high involvement in conflicts, while green indicates a low conflict score. The dynamic nature of the Towers of Hanoi puzzle induces a repetitive structure reflecting the unfolding of the underlying transition function (from left to right). Interestingly, the largest number of conflicts is observed in the "middle" of the problem representation. In fact, the two dark red spots stand for two on/3 fluents. Other indicators than heuristic scores are possible and are made available in a forthcoming visualization tool.

7.2.6   FINE-TUNING

Advanced Boolean constraint technology adds a multitude of degrees of freedom to ASP solving. Currently, *clasp* has more than 60 options, half of which control the search strategy. Although considerable efforts were taken to find default parameters jointly boosting robustness and speed, the default setting still leaves room for drastic improvements on specific benchmarks by fine-tuning the parameters. The question then arises how to deal with this vast "configuration space" and how to conciliate it with the idea of declarative problem solving. Currently, there seems to be no alternative to manual fine-tuning when addressing highly demanding applications.

However, it is crucial to realize that no solver configuration can compensate for a suboptimal encoding. It is thus of the utmost importance to devote all attention to an elaborate encoding before even looking for a good solver configuration.

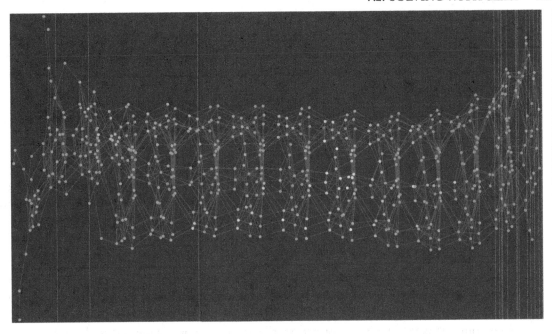

Figure 7.4: Accumulated atom score when solving the Towers of Hanoi problem.

As a rule of thumb, we usually start by investigating the following options:

--heuristic: Try vsids instead of *clasp*'s default berkmin-style heuristic. Sometimes the default heuristic is too time consuming.

--trans-ext: This option determines the treatment of extended rules, such as cardinality and weight rules. We recommend trying the dynamic transformation when the number of Frozen variables is high. Also, it is often advisable when observing very long dynamic nogoods (see also --recursive-str below).

--sat-prepro: SAT-style preprocessing techniques work best on tight programs with few cardinality and weight constraints. We recommend using this option when extended rules are transformed into normal ones (via --trans-ext) and/or when the number of Frozen variables is small.

--restarts: Try aggressive restart policies, like L(uby),256 or the nested policy, or try disabling restarts whenever a problem is deemed to be unsatisfiable.

--save-progress: Progress saving typically works nicely when the average backjump length is high, that is, if the Choices/Conflicts ratio is larger than 10. It usually performs best in combination with aggressive restarts.

--deletion: The need for aggressive nogood deletion relates to the size of a problem, indicated by its number of Variables and Constraints, respectively. On rather large problems, where either number exceeds 100000, we advocate more aggressive nogood deletion, while on smaller ones, where both numbers are below 1000, we use less aggressive settings (combined with a higher restart frequency).

--recursive-str enables *minisat*-like conflict nogood strengthening. We advocate using this whenever one encounters very long dynamic nogoods.

Although fine-tuning may greatly improve the efficiency of *clasp*, it is hard to accomplish for an unpracticed user, and after all, it takes us away from the ideals of declarative problem solving. To this end, we advocate an extension of *clasp*, called *claspfolio*, that maps benchmark features to solver configurations (via machine learning techniques; see Section 7.3.3).

7.2.7   OUTLOOK

Our treatment of *clasp* along with most examples deal with version 2.0.5 (and 2.0.6). Upon completing this book, *clasp* 2.1 was released. Most interesting from a user's perspective is the provision of several pre-configurations of *clasp*. They are engaged via the option --configuration. Many hints from Subsection 7.2.6 are covered by the various configurations. Also, the --help option of *clasp* 2.1 offers an increasing level of detail via an argument. An interesting extension concerns Option --integrate, which now supports different topologies for nogood integration. More information can be found in the corresponding README at potassco.sourceforge.net.

7.3   MORE POTASSCO SYSTEMS

In what follows, we discuss some major extensions of *gringo* and *clasp*. While the first three sections deal with pure solver extensions, the remaining ones deal with ASP systems combining grounding and solving.

Even more Potassco systems are available at potassco.sourceforge.net, notably at potassco.sourceforge.net/labs.html.

7.3.1   *CLASPD*

Many problems in Knowledge Representation and Reasoning have an elevated degree of complexity, calling for expressive solving paradigms being able to capture problems at the second level of the polynomial hierarchy. Such a problem is for instance to decide whether an atom belongs to an inclusion-minimal stable model. They can be addressed by ASP solvers dealing with disjunctive logic programs such as *claspd*.[15]

The actual search for stable models of disjunctive programs is divided into a generating part, providing stable model candidates, and a testing part, verifying the provided candidates. Since

---

[15]Given that *claspd* evolved from an early version of *clasp*, it is about to be re-merged into *clasp*.

both of these tasks can be computationally complex, they are performed by associated inference engines, implemented in *claspd* by feeding the core search module from *clasp* with particular Boolean constraints. While the generator traverses the search space for stable models, communicating its current state through an assignment to the tester, the latter checks for unfounded sets and reports them back in terms of nogoods. An approximate unfounded set detecting procedure is integrated into propagation and thus continuously applied during the generation of stable model candidates. In contrast, exhaustive checks for so-called non-head-cycle-free components are performed only selectively, for instance, if an assignment is total, due to their high computational cost.

7.3.2   *CLASPAR*

claspar is a distributed version of *clasp* constructed for ASP solving on clusters consisting of a multitude of loosely connected computers. To this end, *claspar* is built upon the Message Passing Interface (MPI), realizing communication and data exchange between computing units via message passing. Interestingly, MPI abstracts from the actual hardware and lets us execute our system on clusters as well as multi-processor and/or multi-core machines.

claspar aims at a simple and transparent approach taking advantage of the underlying performance of *clasp*. For this, it relies on simple master-worker architectures, in which each worker consists of an ASP solver along with an attached communication module. The solver is linked to its communication module via an elementary interface requiring only marginal modifications to the solver. All major communication is initiated by the workers' communication modules, exchanging messages with the master in an asynchronous way. The specific communication structure can be configured via the option `--topology`, allowing for flat and more complex hierarchical architectures.

As with multi-threaded *clasp*, its cluster-oriented extension *claspar* supports splitting-based and competition-based search. Moreover, it allows for combining both strategies and having different groups of solvers address distinct search spaces. Also, a portfolio of different *clasp* configurations can be supplied to *claspar* via option `--portfolio-file`. The different configurations are then assigned either randomly or in a round-robin fashion (via `--portfolio-mode`).

Also, *claspar* supports the exchange of nogoods, controlled by two options:

`--nogood-sharing` allows for configuring different strategies for nogood exchange, for instance, depending upon different selection criteria of nogoods to be exchanged and the number of nogoods per communication.

`--nogood-distribution` specifies the communication architecture for nogood exchange. This can be local, depending on the master/worker topology, organized as a hypercube (work nodes are arranged in a hypercube and nogoods are exchanged along the edges), and all to all as well as no exchange at all.

7.3.3   *CLASPFOLIO*

Advanced Boolean constraint technology, as used in *clasp*, is sensitive to parameter configuration. In fact, we are unaware of any true application on which *clasp* is run in its default settings. This parameter sensitivity can be counterbalanced by a portfolio-based approach. To this end, we map a collection of benchmark features onto an element of a portfolio of distinct *clasp* configurations. This mapping is realized via machine learning techniques.

Given a logic program, the goal of *claspfolio* is to automatically select a suitable configuration of *clasp*. In view of the huge configuration space, the attention is limited to some (manually) selected configurations belonging to a portfolio. Each configuration consists of certain *clasp* options. To approximate the behavior of such a configuration, *claspfolio* applies a model-based approach predicting solving performance from particular features of the input. The portfolio used by *claspfolio* contains 25 *clasp* configurations, included because of their complementary performance on a training set. The options of these configurations mainly configure preprocessing, decision heuristics, as well as deletion and restart policies. This provides us with a collection of solving strategies that have turned out to be useful on a range of existing benchmarks. In fact, the hope is that some configuration is (a) well-suited for a user's application and (b) automatically selected by *claspfolio* in view of similarities to the training set.

As shown in Figure 7.5, ASP solving with *claspfolio* consists of four parts. First, the ASP

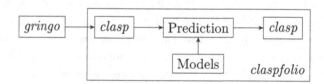

Figure 7.5: Architecture of ASP solver *claspfolio*.

grounder *gringo* instantiates a logic program. Then, *clasp* is used to extract features and possibly even solve (too simple) instances. If the instance was not solved by the initial run of *clasp*, the extracted features are mapped to a score for each configuration in the portfolio. Finally, *clasp* is run for solving, using the configuration with the highest score.

7.3.4   *CLINGO*

The ASP system *clingo* is a monolithic combination of the ASP grounder *gringo* and the ASP solver *clasp* (currently version 1.3). As such, it offers an alternative to using a UNIX pipeline for passing the result of grounding to the solver. Otherwise, *clingo* supports all features and options of *gringo* and *clasp*.

7.3.5 *CLINGCON*

Certain applications are more naturally modeled by mixing Boolean with non-Boolean constructs, for instance, accounting for resources, fine timings, or functions over finite domains. In other words, non-Boolean constructs make sense whenever the involved variables have large domains. This is addressed by the hybrid ASP solver *clingcon*, combining the Boolean modeling capacities of ASP with Constraint Processing (CP). To this end, *clingcon* adopts techniques from the area of SAT Modulo Theories (SMT), combining conflict-driven learning with theory propagation by means of a CP solver. For the latter, we have chosen *gecode* as black box constraint solver.

Although *clingcon*'s solving components, *clasp* and *gecode*, follow the approach of modern SMT solvers, *clingcon* furthermore adheres to the tradition of ASP in supporting a corresponding modeling language via the ASP grounder *gringo*. The resulting tripartite architecture of *clingcon* is depicted in Figure 7.6. *clingcon* extends the input language of *gringo* with theory-specific language

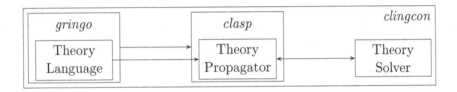

Figure 7.6: Architecture of the hybrid ASP system *clingcon*.

constructs. Just as with regular atoms, the grounding capabilities of *gringo* can be used for dealing with constraint atoms containing first-order variables. This allows for expressing constraints over integer variables, involving arithmetic constraints as well as global constraints and optimization statements. Information about these constraints is directly shared with the theory propagator and in turn the theory solver, here *gecode*. The theory propagator is implemented as (another) post propagator in *clasp*. Theory propagation as such is done through the theory solver.

clingcon follows the lazy approach of advanced SMT solvers by abstracting from the constraints in a specialized theory. The idea is as follows. The ASP solver passes the portion of its (partial) Boolean assignment associated with constraints to a CP solver, which then checks these constraints against its theory via constraint propagation. As a result, it either signals unsatisfiability or, if possible, extends the Boolean assignment by further constraint atoms. For conflict-driven learning within the ASP solver, however, each assigned constraint atom must be justified by a set of (constraint) atoms providing a reason for the underlying inference. Similarly, conflicts occurring within the CP solver have to be justified by such a reason. Both issues are addressed by *clingcon* through dedicated filtering techniques extracting reasons from a current Boolean assignment. This approach also follows the one taken by SMT solvers in letting the ASP solver deal with the atomic, that is, Boolean structure of a problem, while a CP solver addresses the "sub-atomic level" by dealing with the constraints associated with constraint atoms.

To illustrate modeling with *clingcon*, consider the example in Listing 7.7.

Listing 7.7: A hybrid ASP encoding for pouring water into buckets on a balance

```
1   $domain(0..10000).
2   time(0..t).
3   bucket(a).
4   bucket(b).

6   1 { pour(B,T) : bucket(B) } 1 :- time(T), T < t.

8   100 $<= amount(B,T)  :- pour(B,T), T < t.
9   amount(B,T) $<= 300  :- pour(B,T), T < t.
10  amount(B,T) $== 0    :- not pour(B,T), bucket(B),time(T),T < t.

12  volume(B,T+1) $== volume(B,T) $+ amount(B,T) :- bucket(B),time(T),T < t.

14  down(B,T) :- volume(C,T) $< volume(B,T), bucket(B), bucket(C), time(T).
15    up(B,T) :- not down(B,T), bucket(B), time(T).

17  volume(a,0) $== 0.
18  volume(b,0) $== 100.

20  :- up(a,t).
```

This program describes a balance with two buckets, a and b, at each end. According to the cardinality rule in Line 6, we must pour a certain amount of water into exactly one of the buckets at each time point. The amount of added water may vary between 100 and 300 units. The balance is down at one bucket's side, if the bucket contains more water than the other; otherwise, it is up. Initially, bucket a is empty while b contains 100 units. The goal is to find sequences of pour/2 actions making the side of bucket a be down after t time steps.

The program contains regular and constraint atoms. The latter are distinguished via predicate symbols preceded by $. Hence, the amount of water is completely abstracted from the ASP solver and is exclusively handled by the constraint solver. Thus, the capacity can be modeled using any precision and any domain size without increasing the size of the ground program. For example, after instantiation, the ASP solver does not distinguish between the regular atom pour(b,1) and the constraint atom

```
volume(b,2) $== volume(b,1) $+ amount(b,1).
```

It assigns Boolean values to both types of atoms. However, depending on the assigned truth value, the CP solver must assign values to integer variables, volume(b,2), volume(b,1), and amount(b,1), such that the equation evaluates to the assigned truth value.

7.3.6 *ICLINGO*

Many real-world applications, like Automated Planning or Model Checking, have associated PSPACE-decision problems. For instance, the plan existence problem of deterministic planning is PSPACE-complete. But the problem of whether a plan exists whose length is *bounded* by some constant is in NP. In the setting of ASP, such problems can thus be dealt with in a bounded way by considering in turn one problem instance after another, gradually increasing the bound on the solution size.

To illustrate this, let us consider again the Towers of Hanoi puzzle from Section 1.1. This planning problem is solved by the ASP encoding in Listing 1.4 (on Page 6). An answer is usually found by iterative deepening search. That is, one first checks whether the program has a stable model for one move, if not, the same is done for two moves, and so on. Such an approach re-processes all rules parametrized with step variable T multiple times, while the integrity constraint expressing the goal condition in Line 17 of Listing 1.4 is dealt with only once for each bound.

Unlike this, *iclingo* computes stable models in an incremental fashion. The idea is to avoid redundancy by gradually processing the extensions to a problem rather than repeatedly re-processing the entire extended problem. The corresponding architecture is illustrated in Figure 7.7. It is dom-

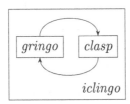

Figure 7.7: Architecture of the incremental ASP system *iclingo*.

inated by a grounder-solver interaction, in which solver results trigger grounding of new program parts.

To capture this, we consider incremental logic programs, consisting of a triple (B, P, Q) of logic programs, among which P and Q contain a (single) parameter k ranging over the natural numbers. In view of this, we also denote P and Q by $P[k]$ and $Q[k]$. The base program B is meant to describe static knowledge, independent of parameter k. The role of P is to capture knowledge accumulating with increasing k, whereas Q is specific for each value of k. Provided all programs are "modularly composable," we are interested in finding a stable model of the program $B \cup \bigcup_{1 \leq j \leq i} P[k/j] \cup Q[k/i]$ for some (minimum) integer $i \geq 1$. In what follows, we write $P[i]$ (or $Q[i]$) rather than $P[k/i]$ (or $Q[k/i]$).

For illustration, let us transform the Towers of Hanoi encoding into an incremental logic program. Clearly, the problem instance in Listing 1.3 as well as the definition of the initial situation belong to the static knowledge in B. As done in Listings 7.8 and 7.9, this is declared by the statement #base.

Listing 7.8: Incremental Towers of Hanoi problem instance (`tohIinc.lp`)

```
1  #base.
2  peg(a;b;c).
3  disk(1..4).
4  init_on(1..4,a).
5  goal_on(1..4,c).
```

In our simple example, the cumulative part consists of all rules possessing step variable T in Listing 1.4. As shown in Listing 7.9, this part is indicated by '#cumulative t', declaring t as the incremental parameter.

Listing 7.9: An incremental ASP encoding of the Towers of Hanoi puzzle (`tohEinc.lp`)

```
1  #base.
2  on(D,P,0)            :- init_on(D,P).

4  #cumulative t.
5  1 { move(D,P,t) : disk(D) : peg(P) } 1.

7  move(D,t)            :- move(D,_,t).
8  on(D,P,t)            :- move(D,P,t).
9  on(D,P,t)            :- on(D,P,t-1), not move(D,t).
10 blocked(D-1,P,t) :- on(D,P,t-1).
11 blocked(D-1,P,t) :- blocked(D,P,t), disk(D).

13 :- move(D,P,t), blocked(D-1,P,t).
14 :- move(D,t), on(D,P,t-1), blocked(D,P,t).
15 :- not 1 { on(D,P,t) } 1, disk(D).

17 #volatile t.
18 :- goal_on(D,P), not on(D,P,t).

20 #hide.
21 #show move/3.
```

Note that t replaces all occurrences of T and makes predicate time/1 obsolete. Finally, the parameter-specific part is indicated by '#volatile t' and applies to the query only.

In order to account for the slice-wise processing of program parts in modeling incremental programs, it is often natural to index dynamic head atoms with the current time stamp t and to refer in body literals to the previous slice via t−1. This convention avoids referring to program parts that have not yet been produced because they are indexed with t+1. In fact, if such atoms are yet undetermined at step t, the addition of rules at step t+1 has no effect on previously produced ground rules.

Incremental programs are solved by the incremental ASP system *iclingo*, built upon the libraries of *gringo* and *clasp*. Unlike the standard proceeding, *iclingo* operates in a "stateful way." That is, it maintains its previous (grounding and solving) state for processing the next program slices. In this way, all components, B, $P[i]$, and $Q[i]$ are dealt with only once, and duplicated work is avoided when increasing i. Launching *iclingo* on the incremental programs in Listings 7.8 and 7.9 yields the (abridged) result in Listing 7.10. Note that we take advantage of *iclingo*'s option --istats to get some insight into the intermediate steps.

Listing 7.10: Incrementally solving the Towers of Hanoi problem

```
$ iclingo tohIinc.lp tohEinc.lp --istats
=============== step 1 ===============

Models   : 0
Time     : 0.000 (g: 0.000, p: 0.000, s: 0.000)
Rules    : 84
Choices  : 0
Conflicts: 0
=============== step 2 ===============

Models   : 0
Time     : 0.000 (g: 0.000, p: 0.000, s: 0.000)
Rules    : 104
Choices  : 0
Conflicts: 0
=============== step 3 ===============

[...]

=============== step 14 ===============

Models   : 0
Time     : 0.000 (g: 0.000, p: 0.000, s: 0.000)
Rules    : 104
Choices  : 87
Conflicts: 27
=============== step 15 ===============
Answer: 1
move(4,b, 1) move(3,c, 2) move(4,c, 3) move(2,b, 4) move(4,a, 5) \
move(3,b, 6) move(4,b, 7) move(1,c, 8) move(4,c, 9) move(3,a,10) \
move(4,a,11) move(2,c,12) move(4,b,13) move(3,c,14) move(4,c,15)

Models   : 1
Time     : 0.000 (g: 0.000, p: 0.000, s: 0.000)
Rules    : 104
Choices  : 80
Conflicts: 24
=============== Summary ===============
SATISFIABLE

Models      : 1+
Total Steps : 15
Time        : 0.000
```

```
Prepare   : 0.000
Prepro.   : 0.000
Solving   : 0.000
```

We observe that *iclingo* initially grounds 84 rules and 104 in each subsequent step. Once the underlying solver is initiated, it is updated fourteen times with new rules. In doing so, it treats fourteen unsatisfiable programs before finding a stable model. Note that whenever a solving step is engaged, it benefits from the information gathered during the previous solving attempts.

Let us take a look under the hood. When processing consecutive program slices, we have to distinguish cumulative and volatile ones. That is, while the ground rules stemming from $P[j]$ are accumulated in the solver, the ones from $Q[j]$ must be discarded for $1 \leq j < i$ when $Q[i]$ is added. This is accomplished by adding to each rule in $Q[j]$ a new body atom α_j, along with rules achieving that α_j holds only at step j (see below). We define the following set of rules for a program Q and a new atom α:

$$Q(\alpha) = \{head(r) \leftarrow body(r) \cup \{\alpha\} \mid r \in Q\}.$$

The addition of new atoms allows us to selectively (de)activate volatile program slices. That is, the fact that programs $Q[j](\alpha_j)$ behave neutrally as long as α_j is unprovable provides us with a handle to control effective program slices. In addition to activating some $Q[j](\alpha_j)$, we also have to deactivate it in subsequent steps. Thus, a solver cannot include α_j persistently as a fact. But rather than explicitly deleting any fact (or rule) previously passed to the solver, we build upon an interface supporting *assumptions*. This trims the required solver interface to only two functions:

- ADD(P) incorporates a ground logic program P into the rule database of the solver;

- SOLVE(L) takes a set L of ground literals and computes the stable models X of the ground program comprised in the solver that satisfy $L^+ \subseteq X$ and $L^- \cap X = \emptyset$.

The literals L passed to SOLVE constitute assumptions, which can be understood as a set $\{\leftarrow \sim a \mid a \in L^+\} \cup \{\leftarrow a \mid a \in L^-\}$ of integrity constraints. However, as regards *clasp*, the crucial difference between integrity constraints and assumptions is that the former give rise to program simplifications affecting internal data structures, while the effect of the latter is temporary, that is, restricted to an invocation of SOLVE. While former assumptions can easily be withdrawn, it would be much more involved to support an explicit deletion of obsolete problem parts.

Semantically, the activation and deactivation of volatile program slices can be captured by choice rules and integrity constraints. A rule like '$\{\alpha_j\} \leftarrow$' nominally permits the unconditional inclusion of α_j in a stable model. Upon the invocation of SOLVE in step j, literal α_j is passed as assumption, so that stable models must necessarily contain α_j. In contrast, in step $j + 1$, an integrity constraint like '$\leftarrow \alpha_j$' can be persistently added to the solver to force α_j to be false. This deactivates all rules from $Q[j]$ in later steps. Notably, *clasp* eliminates such false atoms and rules with false bodies from its data structures, thus automatically deleting a whole obsolete program $Q[j]$.

To make all this more precise, consider Algorithm 7 reflecting the principal functioning of *iclingo*. The algorithm combines grounding with the above solving functions to compute the stable

---

Algorithm 7: ISOLVE

Input : A domain description (B, P, Q).
Output: A non-empty set of stable models.
Internal A grounder Grounder and a solver Solver.
:

1 $i \leftarrow 0$
2 $(P_0, O) \leftarrow$ Grounder.$Ground(B, \emptyset)$
3 Solver.$Add(P_0)$
4 **loop**
5       $i \leftarrow i + 1$
6       $(P_i, O_i) \leftarrow$ Grounder.$Ground(P[i], O)$
7   Solver.$Add(P_i)$
8       $O \leftarrow O \cup O_i$
9       $(Q_i, O_i') \leftarrow$ Grounder.$Ground(Q[i], O)$
10   Solver.$Add(Q_i(\alpha_i) \cup \{\{\alpha_i\} \leftarrow\} \cup \{\leftarrow \alpha_{i-1}\})$
11      $X \leftarrow$ Solver.$Solve(\{\alpha_i\})$
12   **if** $X \neq \emptyset$ **then return** $\{X \setminus \{\alpha_i\} \mid X \in X\}$

---

models of incremental programs (B, P, Q). Programs B, $P[i]$, and $Q[i]$ are then gradually grounded by means of a Grounder and fed into a Solver through function ADD. The state of the grounder is captured via a second argument to the grounding function GROUND that comprises the current set of derivable atoms. In Lines 7 and 10 of Algorithm 7, cumulative and volatile program slices are handled and subsequently solved under (varying) assumption α_i in Line 11. Note that ISOLVE terminates as soon as function SOLVE reports a stable model. Otherwise, if no stable model is found in any step i, ISOLVE (in theory) loops forever. In practice, *iclingo*'s iteration can be controlled via options `--imin`, `--imax`, and `--istop`.

All in all, *iclingo* reduces efforts by avoiding reproducing previous ground rules when extending a program. As regards solving, it reduces redundancy, in particular, if a learning ASP solver such as *clasp* is used, given that previously gathered information on heuristics, conflicts, or loops, respectively, remains available and can thus be continuously exploited. In fact, the latter is configurable via options `--ilearnt` and `--iheuristic` that allow for either keeping or forgetting learned nogoods and heuristic values, respectively.

7.3.7   OCLINGO

All of the above ASP systems are designed for offline usage and thus lack any online capacities. On the other hand, many applications domains involve online data, as agent technology or robotics.

This issue is addressed by the reactive ASP solver *oclingo*, extending *iclingo* with online functionalities. To this end, *oclingo* acts as a server listening on a port (configurable via its --port option upon start-up). Unlike *iclingo*, which terminates after computing a stable model of an incremental logic program, *oclingo* is run on, waiting for client requests. To issue such requests, a separate controller program sends so-called online progressions to *oclingo* and receives stable models to act upon in return. The corresponding architecture is illustrated in Figure 7.8.

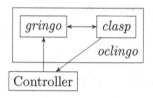

Figure 7.8: Architecture of the reactive ASP system *oclingo*.

For illustrating the usage of *oclingo*, consider a very simple elevator control accepting requests to go to a certain floor whenever it is not already at this floor. At each step, the elevator moves either up or down by one floor. If it reaches a floor for which a request exists, it serves the request and proceeds until its goal to serve all requests is fulfilled. This functionality is specified by the incremental logic program in Listing 7.11.

Listing 7.11: An ASP encoding for simple elevator control (`elevator.lp`)

```
1   #base.
2   floor(1..3).
3   atFloor(1,0).

5   #cumulative t.
6   #external request(F,t) : floor(F).
7   1 { atFloor(F-1;F+1,t) } 1 :- atFloor(F,t-1), floor(F).
8   :- atFloor(F,t), not floor(F).
9   requested(F,t) :- request(F,t),       floor(F), not atFloor(F,t).
10  requested(F,t) :- requested(F,t-1), floor(F), not atFloor(F,t).

12  #volatile t.
13  :- requested(F,t), floor(F).
```

Of particular interest is the declaration preceded by `#external`, delineating the input to the cumulative part provided by putative online progressions. This declaration instructs *oclingo* not to apply any simplifications in view of yet undefined instances of `request(F,t)`.

Launching *oclingo* on Listing 7.11 yields a single stable model containing `atFloor(2,1)`. That is, we make the elevator move one floor and see no pending requests.

In fact, requests are coming from outside the system, and their occurrences cannot be foreseen within an incremental program. If a request to serve floor 3 occurs at time 1, it is passed from the controller to *oclingo* as follows.

```
#step 1.      request(3,1).      #endstep.
```

Here '#step 1' instructs the grounder to (initially) unfold Listing 7.11 until t=1 and to process all rules up to #endstep. Adding the enclosed online progression to the above program yields a stable model containing request(3,1), requested(3,1), atFloor(2,1), and atFloor(3,2), so that the elevator takes two steps to move to the third floor. (Note that this involves unfolding Listing 7.11 until t=2.)

Further language constructs supported by *oclingo*'s controller include #assert, #retract, #forget, as well as extended #volatile statements having a fixed expiration time.

Note that reasoning is driven by successively arriving events. An online progression complements an incremental program as in Listing 7.11 and initiates the subsequent search for a stable model. Afterward, *oclingo* waits for the next request, extending the online progression.

7.4   REFERENCES AND FURTHER READING

Unlike the multitude of ASP solvers, there are only few ASP grounders. Apart from *gringo*, the other two popular grounders are *lparse* (Syrjänen) and (the grounding component of) *dlv* (Leone et al., 2006). While *dlv* processes the grounding result internally (or prints it as text), the *smodels* format produced by *lparse* and *gringo* is compatible with most state-of-the-art ASP solvers. The *smodels* format is detailed in (Syrjänen, Appendix B) and (Janhunen, 2007). An alternative intermediate format is proposed in (Gebser et al., 2008a). Current versions of *gringo* significantly extend the input language of *lparse*. Most notably, they largely eliminate the need for domain predicates. Also, *gringo* has a more general syntax for optimization statements. See (Gebser et al.) for further details. Both *dlv* and *gringo* accept safe programs and tolerate an unrestricted usage of function symbols. We refer the interested reader for details to the literature on *gringo* (Gebser et al., 2007d, 2009b, 2011g) as well as the respective system manuals (Bihlmeyer et al., Gebser et al., Syrjänen).

Since its introduction in Gebser et al. (2007a), *clasp* has become a powerful native ASP solver, offering various reasoning modes that make it an attractive tool for knowledge representation and reasoning. See Gebser et al. (2007a, 2009d, 2012b) on *clasp*'s development up to version 2; its multi-threaded architecture is detailed in Gebser et al. (2012c). Beyond search for stable models of propositional normal programs, *clasp* supports various types of solution enumeration (Gebser et al., 2007b, 2009c) and multi-criteria optimization (Gebser et al., 2011c). Also, more complex modes of reasoning are easily accomplished, like computing the intersection of all optimal models. Moreover, *clasp* features advanced preprocessing techniques, identifying equivalences among (possibly negated) atom and body variables (Gebser et al., 2008c), and a dedicated treatment of cardinality and weight constraints (Gebser et al., 2009a). Specialized variants of *clasp*, like *claspd* (Drescher et al., 2008) for disjunctive ASP solving, *aclasp* using adaptive restarts, or *unclasp* (Andres et al., 2012) using unsatisfiable cores for optimization can be found at potassco. Furthermore, *clasp* can be run as a competitive SAT solver and a solver for Maximum Satisfiability (MaxSAT; Li and Manyà (2009)) and Pseudo-Boolean (PB; Roussel and Manquinho (2009)) constraint satisfaction and/or optimization due to dedicated front-ends supporting the respective input formats. This is witnessed by several first places of *clasp* and *clasp*-based systems at international contests, like ASP, CASC, MISC, PB, and SAT, over the last years.

Apart from the traditional ASP solvers *smodels* (Simons et al., 2002) and *dlv* (Leone et al., 2006), rather different approaches to ASP solving have been pursued, leading to a multitude of ASP solvers. An early implementation of an ASP solver was accomplished by Inoue et al. (1992) based on a model generation theorem prover. Among the contemporary ASP solvers, *smodels-cc* (Ward and Schlipf, 2004) is a first extension of *smodels* with conflict-driven learning; *assat* (Lin and Zhao, 2004), *cmodels* (Giunchiglia et al., 2006), *sag* (Lin et al., 2006), and *sup* (Lierler, 2011) use off-the-shelf SAT solvers (and loop formulas for unfounded set handling); similarly, *pbmodels* (Liu and Truszczyński, 2005) relies on PB solvers for a direct treatment of weight constraints; *lp2sat* (Janhunen, 2006), *lp2diff* (Janhunen et al., 2009), and *lp2mip* (Liu et al., 2012) allow for using off-the-shelf SAT, SMT, and MIP[16] solvers by corresponding translations handling unfounded sets either by a logarithmic translation into SAT or mappings into difference logic or linear constraints, respectively. Combined with a modification of *gringo* as front-end and off-the-shelf SMT and MIP solvers as back-end, the two latter constitute the ASP systems *dingo* and *mingo*. Systems *nomore* (Anger et al., 2001, Konczak et al., 2006) and *nomore++* (Anger et al., 2005) have their roots in Default Logic and use rules or both atoms and bodies as choice points, respectively; *smodels-ie* (Brain and de Vos, 2009) is a modification of *smodels* optimizing cache utilization; *asperix* (Lefèvre and Nicolas, 2009) and *gasp* (Dal Palù et al., 2009) aim at ASP solving while grounding on demand. More information on the diversity of ASP systems can be obtained by consulting the respective ASP competitions (Calimeri et al., 2011, Denecker et al., 2009, Gebser et al., 2007c).

claspd (Drescher et al., 2008) allows for computing stable models of disjunctive logic programs. This can also be accomplished by the ASP solvers *dlv* (Leone et al., 2006), *cmodels* (Giunchiglia et al., 2006), and *gnt* (Janhunen et al., 2006). An early disjunctive ASP solver was presented in Inoue et al. (1992). Ricca et al. (2006) proposed a first backjumping technique for disjunctive logic programming.

claspar was introduced in Ellguth et al. (2009) and further developed in Gebser et al. (2011e), Schneidenbach et al. (2009). It aims at distributing *clasp* on large clusters via the Message Passing Interface (MPI; Gropp et al. (1999)). Earlier approaches to distributed ASP solving include Balduccini et al. (2005), Gressmann et al. (2005, 2006), Pontelli et al. (2003).

claspfolio (Gebser et al., 2011d) is inspired by *satzilla* (Xu et al., 2008). While the latter uses a heterogeneous portfolio of SAT solvers, the former relies exclusively on *clasp* configurations for ASP solving. For learning classifiers, it uses support vector regression (Basak et al., 2007).

---

[16]MIP stands for Mixed Integer Programming.

clingcon (Gebser et al., 2009e) follows the lazy approach of modern SMT solvers (Nieuwenhuis et al., 2006) for combining the ASP solver *clasp* with the CP solver *gecode* (gecode). *clingcon*'s dedicated filtering techniques for extracting reasons and conflicts from Boolean assignments are described in Ostrowski and Schaub (2012). Groundbreaking work on enhancing ASP with CP techniques was conducted in Baselice et al. (2005), Mellarkod and Gelfond (2008), Mellarkod et al. (2008). Balduccini (2009) represents constraint problems via ASP. See Dechter (2003), Rossi et al. (2006) for an introduction to Constraint Processing; and Barrett et al. (2009) on SAT Modulo Theories.

iclingo (Gebser et al., 2008b) combines incremental grounding and solving. The incremental solving interface of *clasp* is similar to the one for incremental SAT solving due to Eén and Sörensson (2003). Meanwhile *iclingo* has been successfully employed in various settings, for instance, for implementing action description languages in *coala* (Gebser et al., 2010a) and PDDL-style planning in *plasp* (Gebser et al., 2011f). Also, we use it as back-end of *fimo* (Gebser et al., 2011j) for implementing a competitive system for finite model generation.

oclingo (Gebser et al., 2011a) extends *iclingo* with capacities for incorporating online data into ASP solving. Gebser et al. (2012a) show how *oclingo* can be used for (window-based) stream reasoning.

A collection of interesting ASP tools due to Tomi Janhunen can be found at Janhunen. See also `potassco.sourceforge.net/labs.html`.

The effectiveness of modern ASP solvers like *clasp* had been impossible without the great progress in Boolean Constraint Solving, mainly conducted in the area of SAT. In view of this, the design and implementation of the SAT solver *minisat* by Eén and Sörensson (2004) deserves particular attention as a primary role model. See also the early approach of *grasp* Marques-Silva and Sakallah (1999), *chaff* (Moskewicz et al., 2001), and *berkmin* (Goldberg and Novikov, 2002). Moreover, the suite of SAT solvers developed by Armin Biere, comprising *lingeling*, *plingeling*, *picosat*, and *precosat* (Biere, 2008, 2010), is a rich source for studying advanced Boolean constraint technology. Finally, the following concepts are of great interest to modern ASP solving: resolution-based preprocessing techniques (implemented in the preprocessor *satelite*; Eén and Biere (2005)); blocked clause elimination (Järvisalo et al., 2010); literal block distance (Audemard and Simon, 2009), that is, the number of distinct decision levels associated with the literals contained in a nogood; heuristics, like *vsids* (Moskewicz et al., 2001), *berkmin* (Goldberg and Novikov, 2002), and *vmtf* (Ryan, 2004); progress saving (Pipatsrisawat and Darwiche, 2007); restart policies (Huang, 2007); dedicated treatment of binary and ternary nogoods (Ryan, 2004); two-watched-literal[17] propagation (Moskewicz et al., 2001); and failed-literal detection (Freeman, 1995).

The first proposal for the ASP input language *ASP-Core-2* is given in Calimeri et al. (2012).

---

[17]Using watch lists for updating constraints is basically an implementation of the well-known *observer design pattern* (Gamma et al., 1994).

CHAPTER 8

Advanced modeling

Modeling in ASP is still an art; it requires craft, experience, and knowledge. Although the resulting ASP encodings are usually quite succinct and easy to understand, crafting an ASP encoding that also leads to the best possible system performance is not yet as obvious as it might seem. This is why ASP modeling is an active and interesting area of current research. To illustrate this, let us begin by conducting an initial case study dealing with the n-queens problem introduced in Section 3.2.

8.1 PIMPING QUEENS

The encoding of this problem in Listing 3.14 on Page 44 was developed according to the fundamental *generate-and-test* methodology of ASP. Starting from a generating rule positioning queens on an $n \times n$ board in a seemingly arbitrary way, we added testing rules eliminating candidate solutions with fewer or more than n queens on the board, and rules excluding two queens on the same row, column, and diagonal.

A first improvement of this encoding is obtained by eliminating symmetric ground rules expressing the same constraint. For example, the integrity constraint

```
:- queen(I,J), queen(I,JJ), J != JJ.
```

in Line 6 of Listing 3.14 gives rise to ground instances

```
:- queen(3,1), queen(3,2).
:- queen(3,2), queen(3,1).
```

both of which prohibit exactly the same placements of queens. This redundancy can be removed by some simple symmetry breaking. In our example, it suffices to replace inequality 'J != JJ' by 'J < JJ'. Globally applying this simple way of symmetry breaking to the encoding in Listing 3.14 yields Listing 8.1.

Listing 8.1: Second attempt at solving the n-queens problem (queens2.1p)

```
1   row(1..n).
2   col(1..n).

4   { queen(I,J) : col(I) : row(J) }.
5    :- not n { queen(I,J) } n.
6    :- queen(I,J), queen(I,JJ),   J < JJ.
7    :- queen(I,J), queen(II,J),   I < II.
```

```
8     :- queen(I,J), queen(II,JJ), I < II, I-J == II-JJ.
9     :- queen(I,J), queen(II,JJ), I < II, I+J == II+JJ.

11  #hide.   #show queen/2.
```

The latter encoding almost halves the number of ground instances obtained from the four integrity constraints. For instance, on the 10-queens problem, the total number of ground rules drops from 2964 to 1494. Despite this reduction, the improved encoding still scales poorly, as witnessed by the 1646904 rules obtained on the 100-queens problem (cf. Table 8.1 at the end of this section).

Analyzing the encoding in Listing 8.1 a bit further reveals that all integrity constraints give rise to a cubic number of ground instances, that is, on the n-queens problem they produce $O(n^3)$ ground rules. This can be drastically reduced by replacing the rule restricting placements in columns, viz. ':- queen(I,J), queen(I,JJ), J < JJ', by

```
:- col(I), not 1 { queen(I,J) } 1.
```

asserting that there is exactly one queen in each column. One rule per column results in $O(n)$ rules (each of size $O(n)$) rather than $O(n^3)$ as before. This is accomplished by virtue of using cardinality constraints providing a compact representation of the pairwise integrity constraints.

Clearly, the same can be done for rows, yielding

```
:- row(J), not 1 { queen(I,J) } 1.
```

Note that the two new rules imply that there is *exactly one* queen per column and row, respectively. Hence, we may actually eliminate the (much weaker) integrity constraint ':- not n { queen(I,J) } n' in Line 5 of Listing 8.1.

Finally, what can we do about the integrity constraints controlling diagonal placements? It fact, the same aggregation through cardinality constraints can be done for the diagonals, once we adopt an enumeration scheme. The idea is to enumerate diagonals in two ways, once from the upper right to the lower left, and similarly from the upper left to the lower right. Let us illustrate this for $n = 4$:

	1	2	3	4
1	1	2	3	4
2	2	3	4	5
3	3	4	5	6
4	4	5	6	7

	1	2	3	4
1	4	3	2	1
2	5	4	3	2
3	6	5	4	3
4	7	6	5	4

A number in the table indicates the respectively numbered diagonal. The two enumeration schemes can be captured by the equations $D = I + J - 1$ and $D = I - J + n$, respectively. For instance, the first equation tells us that diagonal **6** consists of positions $(4, 3)$ and $(3, 4)$, as indicated in bold in the left table. Given both equations, we may replace the rules in Lines 8 and 9 of Listing 8.1, restricting placements in diagonals, by the two following rules:

```
:- D = 1..n*2-1, not { queen(I,J) : D == I-J+n } 1.
:- D = 1..n*2-1, not { queen(I,J) : D == I+J-1 } 1.
```

As above, we thus obtain one rule per diagonal, inducing $O(n)$ ground rules (each of size $O(n)$). The resulting encoding is given in Listing 8.2.[1]

Listing 8.2: Third attempt at solving the n-queens problem (queens3.lp)

```
1  row(1..n).
2  col(1..n).

4  { queen(I,J) : col(I) : row(J) }.

6   :- col(I), not 1 { queen(I,J) } 1.
7   :- row(J), not 1 { queen(I,J) } 1.
8   :- D = 1..n*2-1, not { queen(I,J) : D == I-J+n } 1.
9   :- D = 1..n*2-1, not { queen(I,J) : D == I+J-1 } 1.

11 #hide.  #show queen/2.
```

For 10 and 100 queens, the encoding in Listing 8.2 yields 77 and 797 ground rules, respectively, in contrast to the 1494 and 1646904 rules obtained with the previous encoding in Listing 8.1. Despite the much smaller grounding size, however, the grounding time does not scale as expected. To see this, note that grounding the encoding in Listing 8.2 for 100 queens takes less than a second, while 500 queens require around 25 seconds of grounding time (although only 3997 ground rules are produced).

Further investigations[2] reveal that the last two rules in Listing 8.2 are the source of the problem. In fact, it turns out that during grounding the tests D == I-J+n and D == I+J-1 are repeated over and over. This can be avoided by precalculating both conditions. To this end, we add the rules

```
d1(I,J,I-J+n) :- col(I), row(J).
d2(I,J,I+J-1) :- col(I), row(J).
```

and replace the two conditions D == I-J+n and D == I-J-1 by d1(I,J,D) and d2(I,J,D), respectively. The resulting encoding is given in Listing 8.3.

Listing 8.3: Fourth attempt at solving the n-queens problem (queens4.lp)

```
1  row(1..n).
2  col(1..n).
```

---

[1]We refrained from introducing a domain predicate diag(1..n*2-1) and rather use D = 1..n*2-1 (instead of diag(D)) to avoid unnecessarily blurring grounding size. A construct like I = 1..n can be read as $I \in \{1, \ldots, n\}$.

[2]This can be done with *gringo*'s debug option --verbose, as illustrated in Section 7.1.3.

```
4   { queen(I,J) : col(I) : row(J) }.

6     :- col(I), not 1 { queen(I,J) } 1.
7     :- row(J), not 1 { queen(I,J) } 1.
8     :- D = 1..n*2-1, not { queen(I,J) : d1(I,J,D) } 1.
9     :- D = 1..n*2-1, not { queen(I,J) : d2(I,J,D) } 1.

11   d1(I,J,I-J+n) :- col(I), row(J).
12   d2(I,J,I+J-1) :- col(I), row(J).

14   #hide.   #show queen/2.
```

Although this encoding adds a quadratic number of facts, their computation is straightforward and exploits indexing techniques known from database systems.

Table 8.1 gives some indicative runtime features of the different n-queens encodings. We list results for $n =50$, 100, 500, 1000 with a cut-off at 300 seconds. Each entry gives grounding[3] times, lines of the resulting ground[4] program, and solving[5] time. Times are given in seconds, but only taken from a single run. We observe that the use of cardinality constraints greatly improves

Table 8.1: Experiments contrasting different encodings of the n-queens problem												
n	Listing 3.13			Listing 8.1			Listing 8.2			Listing 8.3		
50	6.27	406804	5.42	0.90	203454	5.29	0.05	397	0.03	0.03	5397	0.03
100	97.67	3293604	—	13.03	1646904	—	0.24	797	0.11	0.09	20797	0.10
500	—	—	—	—	—	—	25.93	3997	5.45	3.21	503997	5.91
1000	—	—	—	—	—	—	195.69	7997	73.16	19.25	2007996	52.88

grounding and solving performance. Moreover, we observe how factoring out relations like d1/3 and d2/3 accelerates grounding despite the quadratic number of additional ground facts.

A common alternative to the previous listings is to merge the generating choice rule in Line 4 with the integrity constraints in Lines 6 and 7 in order to suppress certain invalid selections. The resulting encoding is given in Listing 8.4.

Listing 8.4: Another attempt at solving the n-queens problem (queens5.1p)

```
1   row(1..n).
2   col(1..n).



6    1 { queen(I,J) : row(J) } 1 :- col(I).
```

---

[3]Grounding done with *gringo* 3.0.3.
[4]Obtained using --text.
[5]Solving was done with *clasp* 2.0.5 using --heuristic=vsids.

```
7   1 { queen(I,J) : col(I) } 1 :- row(J).
8   :- D = 1..n*2-1, not { queen(I,J) : d1(I,J,D) } 1.
9   :- D = 1..n*2-1, not { queen(I,J) : d2(I,J,D) } 1.

11  d1(I,J,I-J+n) :- col(I), row(J).
12  d2(I,J,I+J-1) :- col(I), row(J).

14  #hide.  #show queen/2.
```

Solving the 1000-queens problem with the program in Listing 8.4 took 14.24 seconds for grounding and 71.15 seconds for solving. The ground program has one rule less than that obtained from Listing 8.3; also 2000 (body) variables less are obtained from Listing 8.3. Despite this, both programs yield the same number of nogoods. However, this similarity should come as no surprise given that the translation of cardinality constraints in (2.8) on Page 20 turns Listing 8.4 in nearly the same ground program as obtained from Listing 8.3. All in all, the major difference between both encodings thus boils down to the degree of separation among generating and testing parts, and is thus mainly a matter of taste.

8.2   STEMMING BLOCKS

In what follows, we discuss advanced modeling techniques that are often needed in production mode. In doing so, we focus on a modular development carving out various modeling issues. Hence, the individual encodings are no standalone exemplars from the perspective of Knowledge Representation and Reasoning. The chosen issues are representative for many application domains; here they are illustrated via blocks world planning.

A planning problem consists of three parts: an initial situation, a set of actions, and a goal situation. Given such a problem description, a solution is a plan given by a sequence of actions leading from the initial situation to a goal situation.

We consider a single action, move, that allows us to move a block to a location at a certain point in time. A location can be a block or the table. Also, we associate each block with its mass and look for plans minimizing the overall moved mass of blocks. A simple scenario is given in Figure 8.1. In the initial and final situation, we consider nine blocks and a table, yet in different arrangements. Listing 8.5 gives the corresponding problem instance.

Listing 8.5: Initial and goal situation (world.lp)
```
1  step(1..10).

3  block(1..9).

5  init(3,2). init(6,5). init(9,8).
6  init(2,1). init(5,4). init(8,7).
7  init(1,0). init(4,0). init(7,0).
```

```
 9  goal(8,6). goal(5,7).
10  goal(6,4). goal(7,3).
11  goal(4,2). goal(3,9).
12  goal(2,1).

14  mass(7,3).  mass(9,3).
15  mass(5,5).
16  mass(1,7).  mass(3,7).
17  mass(2,11).
18  mass(4,13). mass(6,13).
19  mass(8,15).
```

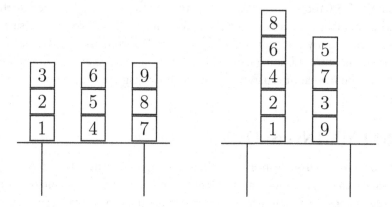

Figure 8.1: Some initial and goal situations for blocks world planning.

The nine blocks are distinguished by predicate block/1 and identified by positive integers. The table is associated with 0. Atom mass(B,M) gives the mass M of block B. We use predicates init/2 and goal/2 to express block locations in the initial and goal situation, respectively. Each state is linked via step/1 to a time point. Note that the initial state is complete, while the goal is only partially defined.

For conducting some empirical evaluations, we handcrafted a parametrized problem instance, world9n.lp, involving $9 \times n$ blocks and $12 \times n$ steps relative to a parameter n.

8.2.1 SEQUENTIAL PLANNING

We start with encodings for sequential planning. A straightforward encoding is given in Listing 8.6.

Listing 8.6: Solving the blocks world problem, initial attempt (blocks0.lp)

```
1  location(0).
2  location(B) :- block(B).
```

```
4  { move(B,L,T) } :- block(B), location(L), step(T), B != L.
5   :- step(T), 2 #count{ move(_,_,T) }.



11  on(B,L,0) :- init(B,L).
12  on(B,L,T) :- move(B,L,T).
13  on(B,L,T) :- on(B,L,T-1), step(T), not no(B,L,T).

15  no(B,L,T) :- location(L), move(B,LL,T), LL != L.

17  blocked(B,T) :- on(_,B,T), block(B), step(T+1).
18   :- move(B,_,T), blocked(B,T-1).
19   :- move(_,B,T), blocked(B,T-1).

21   :- goal(B,L), step(T), not step(T+1), not on(B,L,T).

23  #minimize[ move(B,_,_) = W : mass(B,W) ].

25  #hide.  #show move/3.
```

After defining valid locations in Lines 1 and 2, we generate in Line 4 possible sequences of actions (admitting time points with no action).[6] The integrity constraint in Line 5 eliminates parallel actions. Each state is described in terms of predicate on/3, fixing the positions of blocks at each point in time. In this way, Line 11 captures the initial situation, Line 12 enforces the effect of move actions, and Line 13 progresses positions unless they are changed. A change of a block's location is derived in Line 15. Line 17 tells us which blocks are covered by others. With this, the next two lines eliminate trajectories in which covered blocks are either being moved or serving as target locations. Line 21 ensures that all desired block locations hold in the final state (and thus avoids reoccurring checks at all time points). Finally, the optimization statement in Line 23 selects sequences of actions whose moved objects have a minimum mass. At last, Line 25 projects stable models onto predicate move/3.

Projection   In analogy to Section 8.1, a first source of improvement is the elimination of unnecessary combinatorics inflating the ground program. Such places can be found in Lines 5, 15, and 18, 19 of Listing 8.6. The body of each ground integrity constraint resulting from Line 5 includes a quadratic number of move atoms, although the specific block and location are of no interest. Similarly, the purpose of the atom derived in Line 15 is to indicate that a block has been moved, no matter where to. In this respect, both location L and LL are irrelevant. Similarly, neither constraint in Lines 18 or 19 needs both the moved block as well as its target location.

The refinement of our encoding in Listing 8.7 addresses these issues.

---

[6]Plans with gaps are a special case of plan permutations addressed below.

Listing 8.7: Solving the blocks world problem, first improvement (`blocks1.lp`)

```
1   location(0).
2   location(B) :- block(B).

4   { move(B,L,T) } :- block(B), location(L), step(T), B != L.

6   object(B,T) :- move(B,_,T).
7   target(B,T) :- move(_,B,T).
8    :- step(T), 2 #count{ object(_,T) }.
9    :- step(T), 2 #count{ target(_,T) }.

11  on(B,L,0) :- init(B,L).
12  on(B,L,T) :- move(B,L,T).
13  on(B,L,T) :- on(B,L,T-1), step(T), not object(B,T).


17  blocked(B,T) :- on(_,B,T), block(B), step(T+1).
18   :- object(B,T), blocked(B,T-1).
19   :- target(B,T), blocked(B,T-1).

21   :- goal(B,L), step(T), not step(T+1), not on(B,L,T).

23  #minimize[ object(B,_) = W : mass(B,W) ].

25  #hide.  #show move/3.
```

To this end, Lines 6 and 7 are added to project onto moved blocks and their target locations. The resulting atoms are then used to remodel the spots identified above. First, the single integrity constraint in Line 5 is replaced by Lines 8 and 9. The latter eliminate actions moving more than one object at the same time and moving one object simultaneously to several locations. Note that the resulting ground bodies of both constraints include only a linear number of atoms. The second modification replaces 'not no(B,L,T)' with 'not object(B,T)' in Line 13 and deletes Line 15. A less drastic change, leaving room for further exceptions, would be to leave Line 13 intact and merely replace 'move(B,LL,T), LL != L' with object(B,T) in Line 15. Either change eliminates the quadratic blow-up caused by Line 15 in Listing 8.6. Finally, Lines 18 and 19 of Listing 8.7 are obtained from those in Listing 8.6 by inserting object(B,T) and target(B,T) for their respective definitions. For each step, this yields a linear number of ground integrity constraints rather than the quadratic number obtained from Listing 8.6. The same modification is done to the optimization statement in Line 23. In our simple example, this reduces the number of elements in the ground statement from 810 to 90.

The effect of using projection for reducing combinatorial blow-up can be observed by comparing the performance of blocks0.lp and blocks1.lp when solving our (scalable) problem instance `world9n.lp` for n=1. The four figures in Table 8.2 are obtained by calling *clasp* with option `--stats`.[7] The two times in parentheses give the time spent to find the first model and to prove unsatisfiability, respectively. We observe that our changes have quite a positive effect on the solving behavior. Note that the number of `Rules` does not reflect decreases in number of body literals. Also, it is interesting to note that both runs spend most of their solving time on proving unsatisfiability (in improving obtained solutions).

Table 8.2: Empirically contrasting blocks world encodings

n=1	Rules	Time		Choices	Conflicts
blocks0.lp	14270	20.6	(0.0/20.2)	678074	486795
blocks1.lp	5915	0.9	(0.0/0.8)	35004	24179
blocks2.lp	6899	0.6	(0.0/0.5)	20355	15596
blocks3.lp	8411	0.6	(0.0/0.5)	17940	13352
blocks4.lp	2638	0.0	(0.0/0.0)	105	63
blocks4t.lp	2674	0.0	(0.0/0.0)	105	63
n=4					
blocks4.lp	47980	2.9	(2.1/0.1)	61240	32827
blocks4t.lp	48844	2.1	(1.7/0.0)	44997	24319
n=8					
blocks4.lp	196148	1268.2	(85.2/58.4)	9970769	4071147
blocks4t.lp	199796	81.0	(1.7/0.0)	540772	274277
blocks4P.lp	182559	10.3	(8.9/0.0)	123150	38482
blocks4tP.lp	186207	8.2	(7.34/0.0)	109122	32026
n=16					
blocks4P.lp	738335	195.5	(153.3/36.2)	741518	237702
blocks4tP.lp	753311	157.2	(151.3/0.8)	554611	128681

Symmetry breaking Another source of improvement lies in the elimination of symmetric solution candidates. In fact, in our example, the order of moved blocks is often irrelevant. A good way to accomplish this is to define an order on symmetric solution candidates and then to eliminate inferior candidates in favor of superior ones during search.

Accordingly, our second refinement is obtained by adding the symmetry breaking rules in Listing 8.8 to Listing 8.7. We refer to the resulting encoding as `blocks2.lp`.

Listing 8.8: Solving the blocks world problem, second improvement

```
27  sorted(B+1,T) :- object(B,T), block(B+1), step(T+1).
28  sorted(B+1,T) :- sorted(B,T), block(B+1).
```

[7] All experiments were run with *clasp* 2.0.6.

```
29    :- move(B,L,T), 1 < T, not sorted(B,T-1),
30        not blocked(B;L,T-2), not object(L,T-1).
```

For ordering trajectories, we rely on block numbers. Predicate `sorted/2` gives all blocks inferior to the moved block. For instance, `object(7,3)` induces `sorted(8,3)` and `sorted(9,3)` in our example. The idea is then to eliminate trajectories preferring to move an inferior block to a superior one. In other words, whenever several blocks are movable, the superior one must be picked. This local condition is expressed by the integrity constraint in Lines 29 and 30. Moreover, this constraint enforces that actions are applied as soon as possible.

It is instructive to realize that the definition of `sorted/2` in Listing 8.8 avoids a quadratic blow-up by using linearization. To see this, compare the rules in Lines 27 and 28 with the straightforward definition of `sorted/2`:

```
    sorted(B,T) :- block(B), object(C,T), B > C, step(T+1).
```

While the last rule gives rise to a quadratic number of ground rules, those in Lines 27 and 28 are linear in the number of blocks.

In our initial example, `blocks2.lp` reduces the number of optimal plans from 100 to 5. As an example, consider the following plan eliminated by `blocks2.lp`.

```
        move(9,0,2) move(6,0,3) move(5,0,4) move(3,9,5)
move(4,2,6) move(6,4,7) move(8,6,8) move(7,3,9) move(5,7,10)
```

This plan violates the integrity constraint in Lines 29 and 30 in three ways. First, no action takes place at step 1. Second, block 6 could have been moved at step 2 instead of the inferior block 9. And similarly, block 3 was already movable at step 4 but ignored in favor of inferior block 5. In contrast, the following plan is sanctioned by `blocks2.lp`.

```
    move(9,0,1) move(3,9,2) move(6,0,3) move(5,0,4) move(4,2,5)
    move(6,4,6) move(8,6,7) move(7,3,8) move(5,7,9)
```

The initial move of block 9 is tolerated because it lays the foundation for the second move, and the local nature of our constraint prevents a comparison to the superior block moved at step 3. Other plans start by putting block 6 down.

The effect of symmetry breaking can be read off the third line in Table 8.2. Although the problem size is slightly augmented, we observe an increase in performance. While the decrease in runtime is insignificant, the reduction in terms of choices and conflicts is substantial.

State constraints   A third source of improvement lies in making (redundant) state constraints explicit. To this effect, Lines 32 and 33 of Listing 8.9 ensure that a block can never be at several locations and that several blocks can never be on a single block. Moreover, the rules in Lines 35–37 require that each block must be on a stack of blocks that rests on the table.

Listing 8.9: Solving the blocks world problem, third improvement

```
32   :- block(B), step(T), 2 #count{ on(B,_,T) }.
33   :- block(B), step(T), 2 #count{ on(_,B,T) }.

35   above(B,T) :- on(B,0,T), step(T).
36   above(B,T) :- on(B,L,T), above(L,T).
37   :- block(B), step(T), not above(B,T).
```

We refer to the addition of the rules in Listing 8.9 to Listings 8.7 and 8.8 as `blocks3.1p`. As with our last improvement, the addition of further information removes some choices and conflicts, although this does not translate into a significant gain in runtime in Table 8.2. However, none of the above encodings allows for solving our blocks world instance for n = 2 within an hour.

Background knowledge   A great leap forward is made by incorporating background knowledge about the blocks world domain. For instance, in blocks world planning, individual blocks need never be moved more than twice. That is, they need not be moved if they are already in position, moved once if they can directly be put in position, and twice if they need to be put down before they can be moved to their final destination. And in the latter case, there is at least one block that must be moved first. This kind of background knowledge allows us to significantly reduce the candidate moves in Line 4 (already during grounding). Similarly, it helps us to cut down the elements subject to minimization in Line 23. We refer to the resulting encoding as `blocks4.1p`.

The first change replaces Line 4 by the rules in Listing 8.10. The idea is to group moves in two categories, those directly establishing a goal condition and those putting a block on the table. In addition, the possible moves are constrained by domain constraints extracted from the initial problem specification.

Listing 8.10: Solving the blocks world problem, fourth improvement, Part I

```
39   stay(0).
40   stay(L) :- block(L), init(B,L) : goal(B,L).

42   keep(B,0) :- block(B), not goal(B,L) : block(L).
43   keep(B,L) :- goal(B,L).
44   keep(B,L) :- init(B,L), stay(L), not goal(B,LL) : goal(B,LL) : LL!=L.

46   zero(0).
47   zero(B) :- init(B,L), keep(B,L), zero(L).

49   free(L) :- block(L), not init(B,L) : init(B,L).
50   free(L) :- block(L), init(B,L), fill(B), not zero(B).

52   fill(L) :- zero(L), block(L), not init(B,L) : zero(B).
53   fill(B) :- free(B), keep(B,0).
54   fill(B) :- free(B;L), goal(B,L), fill(L).
```

```
56  once(B) :- keep(B,0), not zero(B).
57  once(B) :- goal(B,L), free(L), fill(L).
58  once(B) :- init(B,L), stay(L), zero(L), not zero(B).

60  { move(B,L,T) } :- goal(B,L), step(T), not zero(B).
61  { move(B,0,T) } :- block(B), step(T), not zero(B), keep(B,0) : once(B).
62    :- move(B,0,T), on(B,0,T-1).
```

In Line 60, a block B is only moved to a location L if this establishes some goal condition goal(B,L). Line 61 handles moves putting blocks down on the table. In addition, both types of moves are restricted via domain predicates zero/1, keep/2, and once/1. While 'not zero(B)' discards blocks that never need to be moved, 'keep(B,0) : once(B)' ensures that the table is among the admissible final locations of B whenever a single move of B is sufficient (and required); otherwise the table may serve as an intermediate location of B. In our simple example, this reduces the number of possible moves from 810 to 100. As regards the instance explored in Table 8.2 (n=1), the passage from blocks3.1p to blocks4.1p yields a reduction from 972 to 144 possible actions. Finally, Line 62 eliminates trajectories in which a block is merely moved around on the table.

The rules in Lines 39 to 58 define domain predicates restricting the instantiation of move/3. All these rules are turned into facts during grounding. The purpose of these predicates is summarized below, accompanied by the facts obtained on Listing 8.5.

- stay/1 identifies locations that need not be cleared.
  stay(0). stay(1). stay(5). stay(8).

- keep/2 gives admissible final locations of blocks.
  keep(1,0). keep(2,1). keep(3,9). keep(4,2). keep(5,7). keep(6,4).
  keep(7,3). keep(8,6). keep(9,0). keep(9,8).

- zero/1 tells us which blocks need never be moved.
  zero(0). zero(1). zero(2).

- free/1 identifies blocks becoming clear upon executing direct moves to goal locations.
  free(2). free(3). free(6). free(8). free(9).

- fill/1 identifies blocks being either in their final position or directly movable there.
  fill(2). fill(3). fill(9).

- once/1 identifies blocks that can be moved directly to their final locations.
  once(3). once(4). once(7). once(9).

Predicates zero/1 and once/1 are central to the encoding because they allow us to identify blocks subject to no move or one move. Below, this is complemented by predicate back/1 isolating blocks that must be moved twice. In our example, we get back(6) as a fact after grounding, leaving us without precise information only about blocks 5 and 8 before solving.

While detecting immobile blocks can be done by matching initial block positions with admissible final locations (cf. Line 47), the identification of one-shot moves is more involved and remains incomplete. Clearly, a block can be put on the table in at most one step (cf. Line 56). Similarly, a block can be directly moved on its final location L, provided the latter is in place (indicated by fill(L)) as well as clear (indicated by free(L)), both of which are checked in Line 57. In our example, this applies to blocks 3 and 9, as represented by facts once(3) and once(9). Finally, a block can be directly moved to its goal position, whenever its underlying stack is stable and needs not be cleared, as expressed in Line 58. In our example, blocks 4 and 7 can directly be moved from the table to their final destinations, as indicated by facts once(4) and once(7).

The interplay of the auxiliary predicates free/1 and fill/1 is interesting to observe. While fill traverses each stack bottom-up, free proceeds top-down. Line 52 identifies the highest block in a stack that should not be moved. Hence, we get fill(2) but not fill(1). Conversely, Line 49 gives the top blocks on each stack, yielding free(3), free(6), and free(9). The latter results in fill(9) via Line 53. As well, Line 54 allows us to derive fill(3) given that block 3 and its final location, block 9, are both free and the final location is readily established, as indicated by fill(9). With fill(3), Rule 50 yields free(2); similarly, we get free(8) in view of fill(9). This tells us that blocks 2 and 9 can be cleared by moving their upper blocks directly on their goal position. This cannot be guaranteed for block 6; hence, free(5) is not obtained.

Another point of interest is the use of conditional literals. For illustration, consider the definition of predicate stay/1 in Line 40, identifying locations that need not be cleared. The conditional literal allows us to cover three cases depending on the goal conditions in Listing 8.5 and the resulting instance of init/2. To see this, let us inspect the (intermediate) ground rules obtained for blocks 1, 2, and 5 after "expanding" the conditional literal in Line 40:

```
stay(1) :- block(1), init(2,1).
stay(2) :- block(2), init(4,2).
stay(5) :- block(5).
```

Body literals with predicate init/2 are obtained whenever the corresponding instance of goal/2 is available. The above rules are then evaluated with respect to Listing 8.5, yielding facts stay(1) and stay(5). In addition, we obtain stay(0) and stay(8).

Similarly, the "expansion" of the conditional literal in Line 44 for (B,L) being (1,0), (2,1), and (3,2) yields the following rules.

```
keep(1,0) :- init(1,0), stay(0).
keep(2,1) :- init(2,1), stay(1).
keep(3,2) :- init(3,2), stay(2), not goal(3,9).
```

The literal not goal(B,LL) is added for all instantiations of B and LL satisfying both goal(B,LL) and LL != L. The above rules result in the facts keep(1,0) and keep(2,1), telling us that the table is an admissible final location for block 1 as is the latter for block 2.

Unlike the above, we get three instances of 'not init(B,L)' from the conditional literal in Line 52 (in view of the extension of predicate zero/1). For example, substituting L by 1 and 2 yields the following ground rules.

```
fill(1) :- zero(1), block(1), not init(0,1), not init(1,1), not init(2,1).
fill(2) :- zero(2), block(2), not init(0,2), not init(1,2), not init(2,2).
```

Also, it is instructive to observe how the conditional literal allows us to identify the highest immobile block in a stack. To see this, note that fill(1) is unobtainable because init(2,1) indicates that it is underneath another immobile block. Unlike this, the body of the second rule is reduced to true yielding fact fill(2).

In Listing 8.11, we exploit further background knowledge for eliminating redundant trajectories.

Listing 8.11: Solving the blocks world problem, fourth improvement, Part II

```
64   done(L,T-1) :- zero(L), step(T), fill(L) : block(L).
65   done(B,T)   :- keep(B,L), on(B,L,T), done(L,T).
66   :- done(B,T), step(T+2), not done(B,T+1).

68   :- object(B,T), done(B,T-1).
69   :- target(L,T), step(T+1), not done(L,T).
70   :- move(B,0,T), goal(B,L), block(L), done(L,T), not blocked(L,T-1).
```

Unlike the above, the principal predicate done/2 is determined during solving; it tells us at which time a block has reached its final destination. done/2 starts in Line 64 to collect placed blocks from the topmost immobile blocks determined statically, and continues in Line 65 with blocks that lie in the current state on an admissible final location. Line 66 ensures that "done blocks" are never undone. Similarly, the remaining rules dismiss moving "done" objects, moves on "undone" targets, and any intermediate storage on the table when direct moves are possible.

Finally, we add further domain knowledge for identifying blocks that must be moved twice. Together with the domain predicates in Listing 8.10, this allows us to significantly reduce the elements subject to minimization. This last change replaces Line 23 by the rules in Listing 8.12; it reduces the number of minimized elements from 108 to 36 (when solving world9n.lp for n=1).

Listing 8.12: Solving the blocks world problem, fourth improvement, Part III

```
72   rank(init,B,B,1)   :- init(B,0).
73   rank(init,A,B,I+1) :- init(B,L), rank(init,A,L,I).

75   rank(goal,B,B,1)   :- keep(B,0).
76   rank(goal,A,B,I+1) :- goal(B,L), rank(goal,A,L,I).

78   back(B) :- rank(init,AI,B,BI), rank(init,AI,L,LI), LI < BI,
79              rank(goal,AG,B,BG), rank(goal,AG,L,LG), LG < BG,
```

```
80                not zero(B).

82  #minimize[ move(B,0,_) = W : mass(B,W)
83                           : not zero(B) : not once(B) : not back(B) ].
```

Predicate rank/4 gives the height of a block in a stack (in the initial and goal situation). Each stack is identified by its bottom block. The idea of the rule in Lines 78–80 is to check whether block B is in both situations above the same block L. A block like B must first be moved away to free L before it can be moved back somewhere above L. Hence, B must be moved twice. In our example, this applies to block 6 only because it is above block 4 in the initial as well as final situation. Hence, we get back(6).

With the help of background knowledge, we were able to identify blocks that need not be moved at all (blocks 1 and 2), once (3, 4, 7, 9), or twice (6). Only blocks 5 and 8 evade this classification (though they must be moved at least once). As a consequence, the contribution of the classified blocks to the objective function is predetermined and need not be included in searching for a minimum. This is accounted for in Line 83 by eliminating all these blocks. The only variable part is whether the unclassified blocks are moved once or twice. Consequently, our objective function only accounts for unclassified blocks that are moved twice. This is done by checking whether they are put down on the table in Line 82. The value of the original objective function in Line 23 can be calculated by summing up the value of the determined moves and the one of the reduced sum in Lines 82 and 83.

To recapitulate, blocks4.lp consists of Listing 8.7 (blocks1.lp) without Lines 4 and 23 along with the rules in Listing 8.8 (yielding blocks2.lp), Listing 8.9 (yielding blocks3.lp), as well as the ones in Listings 8.10, 8.11, and 8.12. Although the effect of incorporating background knowledge is not reflected in the runtime in Table 8.2 for n=1, it is indicative regarding the number of choices and conflicts that drop by three orders of magnitude. While all previous encodings fail to solve our blocks world problem for n=2 within an hour, blocks4.lp still allows for solving the problem for n=8 in about twenty minutes.

A further substantial improvement can be obtained by removing superfluous elements from the optimization statement. In an optimal plan, a block will never be put down twice. Hence, the optimization statement must merely detect whether a block was put down and not when. To accommodate this, we replace the minimization statement in Lines 82 and 83 with the contents of Listing 8.13. We refer to the resulting encoding as blocks4t.lp.

Listing 8.13: Solving the blocks world problem, fifth improvement

```
85  table(B) :- move(B,0,_), not zero(B), not once(B), not back(B).

87  #minimize[ table(B) = W : mass(B,W) ].
```

Although predicate `table/1` can simply be seen as a shorthand for the conditional literal within the minimization statement in Lines 82 and 83, it further reduces the number of minimized elements from 36 to 3 when solving `world9n.lp` for n=1. Scaling up to n=8, this seemingly simple modification leads to a gain of an order of magnitude in runtime and a loss of one order of magnitude in choices and conflicts.

Finally, we mention that the performance can yet be improved by switching to *clasp*'s *vsids*-like heuristics, as shown in Table 8.3. In fact, with `blocks4.lp`, the default setting of *clasp* enumerates

Table 8.3: Empirically contrasting blocks world encodings, using `--heuristic=vsids`

n=8/--heuristic=vsids	Rules	Time		Choices	Conflicts
blocks4.lp	196148	32.5	(1.5/6.7)	233236	120347
blocks4t.lp	199796	7.2	(1.4/0.1)	77660	33840

26 stable models upon finding the optimal one, while the *vsids*-based configuration inspects only 19 models. Comparing Table 8.3 with the corresponding lines in Table 8.2 shows that changing *clasp*'s heuristic takes off another order of magnitude as regards runtime, choices, and conflicts.

8.2.2   PARALLEL PLANNING

Another major improvement can be obtained by allowing for parallel actions. This is because parallel plans give rise to shallower search spaces than sequential ones.

Listing 8.14 gives the parallel version of `blocks3.lp`, originally composed of the rules in Listing 8.9 extended by the ones in Listings 8.7 and 8.8.

Listing 8.14: Solving the blocks world problem, parallel version (`blocks3P.lp`)

```
1   location(0).
2   location(B) :- block(B).

4   { move(B,L,T) } :- block(B), location(L), step(T), B != L.

6   object(B,T) :- move(B,_,T).
7   target(B,T) :- move(_,B,T).



11  on(B,L,0) :- init(B,L).
12  on(B,L,T) :- move(B,L,T).
13  on(B,L,T) :- on(B,L,T-1), step(T), not object(B,T).



17  blocked(B,T) :- on(_,B,T), block(B), step(T+1).
```

```
18    :- object(B,T), blocked(B,T-1).
19    :- target(B,T), blocked(B,T-1).

21    :- goal(B,L), step(T), not step(T+1), not on(B,L,T).

23    #minimize[ object(B,_) = W : mass(B,W) ].

25    #hide.   #show move/3.



29    :- move(B,L,T), 1 < T,
30       not blocked(B;L,T-2), not object(L,T).

32    :- block(B), step(T), 2 #count{ on(B,_,T) }.
33    :- block(B), step(T), 2 #count{ on(_,B,T) }.

35    above(B,T) :- on(B,0,T), step(T).
36    above(B,T) :- on(B,L,T), above(L,T).
37    :- block(B), step(T), not above(B,T).
```

Listing 8.14, or blocks3P.lp, is obtained from blocks3.lp via two changes. First, Lines 8 and 9 are removed given that they prevent parallel actions. Second, the symmetry breaking rules in Listing 8.8 are reduced to Lines 29 and 30 above. That is, the definition of sorted/2 is removed and the occurrence of 'not sorted(B,T-1)' is deleted in Line 29. Also, 'not object(L,T-1)' is turned into 'not object(L,T)' in Line 30. The remaining integrity constraint enforces that actions are applied as soon as possible.

Note that the resulting encoding is rather liberal and for instance allows us to move a block on another that is also about to be moved. The consistency of the respective states is ensured by the state constraints in Lines 32 to 37.

The search for parallel plans is no silver bullet. In fact, the search space induced by blocks3P.lp is much less constrained than that of blocks3.lp. The former results in a performance being orders of magnitude worse than obtained with the latter. For instance, it takes more than an hour to solve world9n.lp for n=1 with blocks3P.lp, while some milliseconds suffice with blocks3.lp. However, this changes once we take advantage of background knowledge again. The encodings blocks4P.lp and blocks4tP.lp are obtained from blocks3P.lp as described in Section 8.2.1. Looking at Table 8.2, we observe how smoothly the parallel versions scale once the search space has been streamlined by additional domain knowledge.

Finally, we mention that changing the heuristic also yields an improvement in the parallel setting. For instance, a *vsids*-based configuration of *clasp* returns a (proven) optimal model of blocks4tP.lp and world9n.lp for n=16 after 66.8 seconds, 150188 choices, and 67135 conflicts.

All in all, our elaboration upon assorted encoding techniques by means of blocks world planning demonstrates that the best performance is obtained by combining various techniques rather than applying a single one. Such encoding techniques include "syntactic" ones, like projection, symmetry breaking, linearization, and the addition of redundant constraints. Beyond that, "semantic" tuning by means of background knowledge and viewpoint shifts (for instance, from sequential to parallel planning) can often boost the performance by orders of magnitude.

8.3   SPEEDING SALESMEN

We have already seen in the previous section how a more careful formulation of an objective function may improve the solving process. So far, however, the elements of an optimization statement have been regarded as being independent of each other. Often enough, this is not the case, and interdependencies can be exploited provided they are made available to the solver.

To illustrate this, let us reconsider the traveling salesperson problem from Section 3.3. Lines 1 to 11 in Listing 8.15 correspond to the combined programs of Listing 3.16 (ham.lp on Page 46) and 3.18 (min.lp), except that Rule 4 in Listing 3.16, viz. 'reached(Y) :- cycle(1,Y)', is replaced in Listing 8.15 to deal with graphs lacking a node labeled 1. We refer to the program consisting of Lines 1 to 11 in Listing 8.15 as tsp.lp.

Listing 8.15: Traveling salesperson, simple encoding (tsp.lp)

```
1   1 { cycle(X,Y) : edge(X,Y) } 1 :- node(X).
2   1 { cycle(X,Y) : edge(X,Y) } 1 :- node(Y).

4   reached(X) :- X = #min[ node(Y) = Y ].
5   reached(Y) :- cycle(X,Y), reached(X).

7   :- node(Y), not reached(Y).

9   #hide.  #show cycle/2.

11  #minimize[ cycle(X,Y) = C : cost(X,Y,C) ].
```

For an example, we use the graph in Figure 3.8 on Page 46 along with its factual representation in Listings 3.1 (graph.lp) and 3.15 (costs.lp).

To put the above mentioned idea into practice, we replace Line 11 in Listing 8.15 by the rules in Listing 8.16. We call the obtained program tspA.lp.

Listing 8.16: Traveling salesperson, advanced minimization

```
13  outcost(X,C) :- cost(X,_,C).

15  order(X,C1,C2) :- outcost(X,C1;C2), C1 < C2,
16                    C <= C1 : outcost(X,C) : C < C2.

18  penalty(X,C1,C2-C1) :- order(X,C1,C2), cycle(X,Y), cost(X,Y,C2).
```

```
19  penalty(X,C1,C2-C1) :- order(X,C1,C2), penalty(X,C2,_).

21  #minimize[ penalty(_,_,C) = C ].
```

To begin with, we exploit the functional dependencies expressed in Lines 1 and 2 and associate with each node the cost of traversing it. We choose in Line 13 the costs of outgoing edges, though incoming ones work just the same. For instance, node 1 has outcost(1,1), outcost(1,2), and outcost(1,3). In turn, we order the costs of each node via predicate order/3 in Lines 15/16 (using linearization to avoid a quadratic blow-up). Here, the conditional literal 'C <= C1 : outcost(X,C) : C < C2' ensures that C1 is the next smaller element to C2. In this way, we get for node 1 the facts order(1,1,2) and order(1,2,3). Now, depending on the chosen outgoing edge a penalty is imposed. This penalty follows the order of costs associated with each edge. That is, the edge with the lowest cost is free. The next one calls for a penalty consisting of the difference in cost to the next lower one, etc. The overall penalty for "leaving" a node is given by the sum of penalties starting from the chosen outgoing edge down to the edge with the lowest cost. As an example, consider the instantiation of Lines 18 and 19 for node 1.

```
penalty(1,2,1) :- cycle(1,3).
penalty(1,1,1) :- cycle(1,2).

penalty(1,1,1) :- penalty(1,2,1).
```

No penalty is caused for cycle(1,4), while cycle(1,3) and cycle(1,2) give rise to a penalty of 1. However, choosing the former edge also adds the penalty caused by the latter.

Now, instead of searching for an optimal assignment to the independent cycle/2 atoms in Line 11, the minimization statement in Line 21 relies on penalty/3 atoms that are inferentially related via Rule 19. (The values of both objective functions differ but can be converted into each other.)

To provide an idea on the computational impact of the rules in Listing 8.16, we ran both encodings on a series of so-called clumpy graphs. While tspA.lp completed 56 out of 77 problems within three minutes, tsp.lp solved only 39. In order to get a more detailed view, we give in Table 8.4 some information on the most difficult problems per class (no instance was solved by tsp.lp in Class 11). A graph labeled n/m belongs to the class of clumpy graphs with n clumps of n nodes each; m is an identifier. The figures in Table 8.4 are obtained by calling *clasp* with option --stats.[8] Although the ground program obtained from both encodings is nearly the same, the inferential dependencies among the elements of the objective function have a significant impact ranging from one to three orders of magnitude. The rules in Listing 8.16 do not only reduce the number of conflicts and choices but also decrease the ratio between them.

The previous example has illustrated the advantage of a knowledgeable optimization statement over an uninformed one, whose elements are prone to plain combinatorics. Directing the optimization

---

[8] All experiments were run with *clasp* 2.0.6 using --heuristic=vsids.

Table 8.4: Empirically contrasting traveling salesperson encodings

Graph	tsp.lp				tspA.lp			
	Rules	Time	Choices	Conflicts	Rules	Time	Choices	Conflicts
06/05	1104	2.0	228291	210797	1628	0.2	18091	15875
07/05	1556	126.4	12566913	11069578	2343	1.2	92975	77528
08/10	1655	25.0	2210992	2110187	2410	1.4	98178	85412
09/11	2087	164.4	14434424	13862523	3048	1.0	81873	65024
10/05	2527	135.2	9606607	8972659	3662	2.1	120243	99293
12/03	3573	164.7	8868378	8718742	5129	1.4	81583	68359

process by providing knowledge about its entities is usually highly beneficial and a key to many application domains.

8.4   VADE MECUM GUM

Although ASP offers an easy declarative modeling language, scalability is not for free. While we are relieved from telling a system how to solve a problem, scalability calls for representing a problem such that a solver can draw advantage from its representation.

In what follows, let us give some hints on scalable modeling based on our experience.

The foundation of each ASP encoding is laid by the *generate-and-test* methodology presented in Section 3.2. All following refinements adhere to this principled approach.

A rough guideline for creating a robust encoding is the following one.

1. Create a working encoding following the *generate-and-test* methodology.

2. Revise the encoding until all of the following questions are answered with "No!"

   (a) Do you need to modify the encoding if the facts change?

   (b) Are some variables insignificant (and functionally independent)?

   (c) Can there be (many) identical ground rules?

   (d) Can you use aggregates (unless they hamper learning)?

   (e) Do you enumerate pairs of values (to test uniqueness)?

   (f) Do you assign dynamic aggregate values (to check a fixed bound)?

   (g) Do you admit (obvious) symmetric solutions?

   (h) Do you have additional domain knowledge simplifying the problem?

   (i) Are you aware of anything else that, if encoded, would reduce grounding and/or solving efforts?

In addition, one may consider revising the fact format as well, whenever it makes encoding painful (for instance, abusing grounding for involved calculations).

During the development of an encoding, two kinds of errors may occur. On the one hand, syntactic errors might be signaled by the grounder. To address them, it is usually sufficient to follow the grounder's error messages and fix them at the source. On the other hand, semantic errors are (most likely) caused by a mismatch between the user's intention and the encoding at hand. Here are some ways to identify such errors early on.

1. Develop and test incrementally.

   - Prepare toy instances with "interesting features."
   - Build the encoding bottom-up and verify additions (e.g., new predicates).

2. Compare the encoded to the intended meaning.

   - Check whether the grounding fits (use *gringo* option `--text`).
   - If stable models are unintended, investigate conditions that fail to hold.
   - If stable models are missing, examine integrity constraints (and add heads).

3. Use ASP-oriented Software Engineering tools.

Once a running encoding is built, one may have to overcome some performance bottlenecks. As regards grounding, it is advisable to

1. monitor the *time* spent and the output *size* via

   - system tools (e.g., `time(gringo <file> | wc)`) or
   - profiling info (as discussed in Section 7.1.3), and

2. once identified, reformulate "critical" logic program parts.

Regarding solving, it is worthwhile to

1. check solving statistics (as discussed in Section 7.2.5) and

2. if great search efforts (`Conflicts`, `Choices`, `Restarts`) are detected, then

   - try auto-configuration (offered by *claspfolio*),
   - try manual fine-tuning (requires expert knowledge), or
   - if possible, reformulate the problem or add domain knowledge ("redundant" constraints) to help the solver.

And, after all, take a breath and enjoy ASP!

8.5   REFERENCES AND FURTHER READING

See Section 3.4 for literature references and further reading on modeling.

Blocks world planning in ASP is discussed in Lifschitz (1999, 2002). Slaney and Thiébaux (2001) provide a detailed analysis of the blocks world domain along with dedicated strategies. General strategies for parallel planning can be found in Rintanen et al. (2006).

Software Engineering tools for ASP are summarized in Chapter 9.

CHAPTER 9

Conclusions

Answer Set Programming has made tremendous progress since the turn of the millennium.

The advancement of ASP is greatly boosted by the extraordinary effectiveness of modern Boolean constraint technology, pioneered in the area of Satisfiability Testing. The distinguishing feature of ASP lies in its versatility. ASP cannot only be uniformly used for MaxSAT, PB, 2QBF, SAT, and dedicated SMT solving, but moreover adds novel problem solving capacities. For example, unfounded-set-based inferences enable closed world reasoning and offer enhanced modeling capacities, as witnessed by the ease of capturing reachability. Moreover, complex forms of reasoning, such as solution enumeration and projection, intersection and union, as well as optimization can be combined in ASP in various ways.

However, the major asset of ASP lies in its easy yet powerful modeling language, providing an elaboration-tolerant tool for Knowledge Representation and Reasoning. This is greatly supported by highly effective grounding systems that rely on (deductive) database technology. Meanwhile, these modeling capacities are also available for SAT and certain SMT solvers via tools like *lp2sat* and *lp2diff*.

9.1 OMISSIONS

There is much more to ASP than we could cover in this book. To compensate this, let us sketch some of ASP's subareas that we recklessly neglected in our treatment.

Applications Foremost, ASP has been applied in many distinct areas. Among them, we find Automated Planning (Dimopoulos et al., 1997), code optimization (Brain et al., 2006), composition of Renaissance music (Boenn et al., 2008), database integration (Leone et al., 2005), decision support for NASA shuttle controllers (Nogueira et al., 2001), Model Checking (Heljanko and Niemelä, 2003, Liu et al., 1998), product configuration (Soininen and Niemelä, 1999), Robotics (Chen et al., 2009, 2010, Erdem et al., 2011), System Biology (Erdem and Türe, 2008, Gebser et al., 2010b, 2011k), System Synthesis (Ishebabi et al., 2009), (industrial) team-building (Grasso et al., 2010), and many more.

Intuitionistic foundations The exploration of ASP's constructive nature was laid bare by Pearce in (1994, 1996). This line of research led to the development of Equilibrium Logic (Pearce, 2006) that can be regarded as ASP's underlying logical formalism.

Equivalence In fact, Equilibrium Logic was used by Lifschitz et al. (2001) to capture the notion of strong equivalence of logic programs under stable models semantics. Two programs are strongly equivalent if they yield the same stable models, no matter which set of rules is added to both of them. This concept guarantees the substitution of equivalent program parts. A similar concept is uniform equivalence capturing the equivalence of logic programs relative to varying sets of facts (Eiter and Fink, 2003). See Woltran (2011) for a brief survey.

Modularity The composition of logic programs must be guided by properties guaranteeing a well-defined semantics of the resulting program. For example, splitting provides a recipe for dividing a program into parts (Lifschitz and Turner, 1994). The atoms in one part's stable models can in turn be used as facts when computing the stable models of the other (that gives the stable models of the overall program). A more general approach is offered by the module theory developed by Oikarinen and Janhunen (2006). Roughly, two programs can be composed if their union does not yield any cycles in their joint positive atom dependency graph. Extensions of this concept are central semantic underpinnings to incremental and reactive ASP (Gebser et al., 2008b, 2011a).

Change The update of logic programs is much more delicate than in propositional logic because of nonmonotonicity. This has led to a variety of distinct approaches to updating logic programs, like Alferes et al. (2002), Delgrande et al. (2007), Eiter et al. (2002), Leite (2003), Sakama and Inoue (1999), Zhang and Foo (2005). Semantic approaches to revision, merging, and update were studied in Delgrande et al. (2012), Slota and Leite (2010).

Preferences We have already seen in Section 2.3.3 how weighted literals along with optimization statements can be used to express preferences. Many more forms of preferences have been developed in ASP, ranging from preferences over rules (Brewka and Eiter, 1999, Delgrande et al., 2003, Schaub and Wang, 2003) to complex optimization techniques based on summation, inclusion, or Pareto-efficiency (Brewka, 2002, Gebser et al., 2011h, Sakama and Inoue, 2000). See Delgrande et al. (2004) for an overview.

Software engineering The pragmatic aspects of software engineering become more and more important to ASP in view of its increasing range of applications. Among others, this involves effective development tools, including editors and debuggers, as well as the dissemination of (open-source) tools and libraries connecting ASP to other computing paradigms. First development environments are described in Febbraro et al. (2011), Oetsch et al. (2011). Preliminary tools for visualization are sketched in Bösel et al. (2004), Cliffe et al. (2008). Interfaces to host languages, databases, and ontologies are presented in Eiter et al. (2006, 2008), Ricca (2003), Ricca et al. (2009), Terracina et al. (2008). In *gringo*, the latter is addressed in a generic way by means of its integrated scripting language (see Section 7.1.2).

Moreover, traditional debugging techniques often do not even apply to ASP because of its strict separation of logic and control. In other words, they fail because ASP lacks a procedural semantics that could be subject to conventional debugging and tracing. This "curse of declarativity" is well recognized within the ASP community and addressed within a dedicated workshop series (de Vos and Schaub, 2007, 2009). Approaches to debugging can be found in (Brain and de Vos, 2005, Brain et al., 2007, Gebser et al., 2008d, Pontelli and Son, 2006, Syrjänen, 2006).

First-order ASP So far, our treatment of first-order variables has been confined to variables over terms, in other words, stable models were regarded as Herbrand interpretations. True quantification over arbitrary domains can either be studied in Equilibrium Logic (Pearce, 2006) or dedicated frameworks generalizing ASP (Ferraris et al., 2007).

9.2 CHALLENGES

Despite the tremendous progress made in ASP over the last decade, the field still faces significant research challenges. In what remains, let us give a personal account of what needs to be done.

Dynamic systems and reactive reasoning Many real-world applications take place in a dynamic setting. Looking at the success stories of SAT in Automated Planning (Kautz and Selman, 1992) and Model Checking (Clarke et al., 2001), we observe that both deal with dynamic systems. Although similar attempts were made in ASP, they did not have the same impact. On the one hand, this is due to the fact that both domains have dedicated problem description languages, diminishing the need for an elaboration-tolerant modeling language. On the other hand, ASP has neglected research on design patterns aiming at search space reductions in dynamic settings (see Rintanen (2009)) and lags behind in the development of incremental grounding and solving techniques (cf. Section 7.3.6 or Eén and Sörensson (2003)). The first challenge is thus to consolidate ASP's general-purpose modeling capacities with such dedicated techniques for dynamic domains.

Moreover, the rapid advance of Internet and Sensor technology increases this need in view of the emergence of dynamic data streams, like web logs, mobile locations, or traffic data. However, the resulting applications do not only require the ability to reason about but also within a dynamic domain. The issuing challenge consists in developing ASP technology that is capable of dealing with online data streams. This includes grounding and solving technology for reactive ASP along with novel modeling capacities dealing with emerging as well as expiring data in a seamless way. A first step in this direction is done in Gebser et al. (2011a, 2012a).

Optimization and preference handling Another commonality of many real-world applications is their interest in optimal solutions.

Although quantitative preferences as presented in Section 2.3 have been a part of ASP systems from the very beginning, the underlying computational methods have not progressed much. All major ASP systems implement optimization with branch and bound algorithms focusing on an upper bound. An exception to this is the prototypical ASP solver *unclasp* whose optimization is based on computing unsatisfiable cores, as developed in the area of MaxSAT. Similarly, there is little work in ASP on incorporating lower bounds or non-uniform bound manipulation. Hence, it is an important future challenge to conceive ways to incorporate advanced optimization techniques from neighboring areas like MaxSAT and even more remote fields like Combinatorial Optimization.

Another challenge lies in the extension of ASP's modeling capacities with language fragments capturing qualitative and quantitative preferences. The possibility to uniformly express both types of preferences would provide another asset of ASP as a powerful knowledge representation tool. A promising implementation platform for this endeavor is offered by meta programming; see also Eiter et al. (2003), Gebser et al. (2011h).

True declarativity Unlike traditional (logic) programming, ASP has succeeded in strictly separating logic from control. Despite this, it is arguably not fully declarative because the way we encode a problem still influences the performance of finding its solution. In other words, two equivalent encodings may result in a significantly different runtime behavior in terms of grounding and/or solving. Although the ideal of declarativity appears to be unattainable, ASP still leaves plenty of room for improvement.

On the one hand, we have seen that ASP modeling is still an art, requiring craft, experience, and knowledge. While ASP encodings are usually quite succinct and easy to understand, designing an encoding that is both elegant and scalable is still not as obvious as it might seem. For addressing this shortcoming, we need methods for automated program optimization, foremost on the non-ground level in order to avoid combinatorial blow-ups in grounding. A role model for this is database query optimization. First attempts at non-ground preprocessing were made in Faber et al. (1999), Gebser et al. (2011b).

On the other hand, advanced Boolean Constraint Solving is sensitive to parameter tuning. Clearly, this carries over to modern ASP solving. For addressing this shortcoming, we have to understand how structural properties of ASP encodings (and instances) influence parameters, steering the search for stable models. As a simple example, consider the relation between tight programs and unfounded set checking. One way to overcome the sensitivity to parameters is to use classifiers from machine learning as done in *claspfolio*. But this cannot relieve us from the task of developing a systematic understanding on how a problem's structure influences the search for its solutions.

9.3   ARMING TWEETY WITH JET ENGINES

Beginning in Nonmonotonic Reasoning with phenomenon-oriented studies in commonsense reasoning in the eighties, ASP has evolved into an attractive declarative problem solving paradigm. The unique pairing of declarativeness and performance allows for concentrating on an actual problem, rather than a smart way of implementing it. ASP's development has also led to evolving problem scenarios, beginning with the (in)famous Tweety examples, to artificial combinatorial problems, up to first success stories in industrial application domains. However, the approach of ASP is not only suitable for the practitioner solving a problem at hand but also for disseminating many basic AI techniques through teaching their (executable) formalization in ASP. The ASP community can be proud of having taught Tweety how to fly. But arming Tweety with jet engines is not enough. Despite its increasing popularity, ASP cannot yet be regarded as an established technology, matching the needs for a widely used problem solving paradigm. But even though there is still a long way to go before having established ASP among the standard technologies in Informatics, its future is bright and conceals many interesting research challenges.

APPENDIX A

ASP in a nutshell

This appendix provides a self-contained introduction to the syntax and semantics of ASP's core language; it is meant to serve as a compact reference. The definition of stable models provided below applies to logic programs containing weight constraints under "choice semantics" (as used in Section 2.3.2), while additionally allowing for disjunctions under minimal-model semantics (with respect to a reduct).

A rule r is of the following form:

$$H \leftarrow B_1, \ldots, B_m, \sim B_{m+1}, \ldots, \sim B_n.$$

By $head(r) = H$ and $body(r) = \{B_1, \ldots, B_m, \sim B_{m+1}, \ldots, \sim B_n\}$, we denote the *head* and the *body* of r, respectively, where $\sim$ stands for default negation. The head H is a disjunction $a_1 \vee \cdots \vee a_k$ over *atoms* $a_1, \ldots, a_k$, belonging to some alphabet $\mathcal{A}$, or a *weight constraint* of the form '$l \, \#sum[\ell_1 = w_1, \ldots, \ell_k = w_k] u$'. In the latter, $\ell_i = a_i$ or $\ell_i = \sim a_i$ is a *literal* and w_i a *non-negative*[1] integer *weight* for $a_i \in \mathcal{A}$ and $1 \leq i \leq k$; l and u are integers providing a lower and an upper bound. Either or both of l and u can be omitted, in which case they are identified with the (trivial) bounds 0 and ∞, respectively. A rule r such that $head(r) = \bot$ (that is, H is the empty disjunction) is an *integrity constraint*. Each body component B_i is either an atom or a weight constraint for $1 \leq i \leq n$. If $body(r) = \emptyset$, r is called a *fact*, and we skip $\leftarrow$ when writing facts. For a set $\{B_1, \ldots, B_m, \sim B_{m+1}, \ldots, \sim B_n\}$, a disjunction $a_1 \vee \cdots \vee a_k$, and a weight constraint '$l \, \#sum[\ell_1 = w_1, \ldots, \ell_k = w_k] u$', we let

$$\{B_1, \ldots, B_m, \sim B_{m+1}, \ldots, \sim B_n\}^+ = \{B_1, \ldots, B_m\},$$
$$(a_1 \vee \cdots \vee a_k)^+ = \{a_1, \ldots, a_k\}, \text{ and}$$
$$(l \, \#sum[\ell_1 = w_1, \ldots, \ell_k = w_k] u)^+ = [\ell_i = w_i \mid 1 \leq i \leq k, \ell_i \in \mathcal{A}].$$

Note that the elements of a weight constraint form a multiset, possibly containing duplicates. For some $S = \{a_1, \ldots, a_k\}$ or $S = [a_1 = w_1, \ldots, a_k = w_k]$, we define $atom(S) = \{a_1, \ldots, a_k\}$.

An *interpretation* is represented by the set $X \subseteq \mathcal{A}$ of its entailed atoms. The satisfaction relation $\models$ on rules like r is inductively defined as follows.

1. $X \models \sim B$ if $X \not\models B$,

2. $X \models (a_1 \vee \cdots \vee a_k)$ if $\{a_1, \ldots, a_k\} \cap X \neq \emptyset$,

[1] In view of the discussion on Page 22, we restrict ourselves to non-negative integers.

3. $X \models (l \,\#sum[\ell_1 = w_1, \ldots, \ell_k = w_k] \, u)$ if $l \le \sum_{1 \le i \le k, X \models \ell_i} w_i \le u$,

4. $X \models body(r)$ if $X \models \ell$ for all $\ell \in body(r)$, and

5. $X \models r$ if $X \models head(r)$ or $X \not\models body(r)$.

A *logic program* P is a set of rules r, and X is a *model* of P if $X \models r$ for every $r \in P$. The reduct of the head H of a rule r with respect to X is

$$
\begin{aligned}
H^X &= \{a_1 \vee \cdots \vee a_k\} & \text{if} \quad H &= a_1 \vee \cdots \vee a_k, \text{ and} \\
H^X &= atom(H^+) \cap X & \text{if} \quad H &= l \,\#sum[\ell_1 = w_1, \ldots, \ell_k = w_k] \, u.
\end{aligned}
$$

Furthermore, the reduct of some (positive) body element $B \in body(r)^+$ is

$$
\begin{aligned}
B^X &= B & \text{if} \quad B &\in \mathcal{A}, \text{ and} \\
B^X &= \left(l - \sum_{1 \le i \le k, \ell_i = \sim a_i, a_i \notin X} w_i\right) \#sum \, B^+ & \text{if} \quad B &= l \,\#sum[\ell_1 = w_1, \ldots, \ell_k = w_k] \, u.
\end{aligned}
$$

The *reduct* of P with respect to X is the following logic program:

$$
\begin{aligned}
P^X = \{ H &\leftarrow B_1^X, \ldots, B_m^X \mid \\
&r \in P, X \models body(r), H \in head(r)^X, body(r)^+ = \{B_1, \ldots, B_m\}\}.
\end{aligned}
$$

That is, for all rules $r \in P$ whose bodies are satisfied with respect to X, the reduct is obtained by replacing weight constraints in heads with individual atoms belonging to X and by eliminating negative components in bodies, where lower bounds of residual weight constraints (with trivial upper bounds) are reduced accordingly. Finally, X is a *stable model* of P if X is a model of P such that no proper subset of X is a model of P^X. In view of the latter condition, note that a stable model is a *minimal* model of its own reduct.

As in Section 2.3.3, *minimize statements* are of the following form:

$$
\#minimize[\ell_1 = w_1 @ p_1, \ldots, \ell_k = w_k @ p_k]. \tag{A.1}
$$

As with weight constraints, every ℓ_i is a literal and every w_i an integer weight for $1 \le i \le k$, while p_i additionally provides an integer *priority level*. Priorities allow for representing a sequence of lexicographically ordered minimize objectives, where greater levels are more significant than smaller ones. A minimize statement distinguishes optimal stable models of a program P in the following way. For any $X \subseteq \mathcal{A}$ and integer p, let Σ_p^X denote the sum of weights w over all occurrences of weighted literals $\ell = w @ p$ in (A.1) such that $X \models \ell$. A stable model X of P is dominated if there is a stable model Y of P such that $\Sigma_p^Y < \Sigma_p^X$ and $\Sigma_{p'}^Y = \Sigma_{p'}^X$ for all $p' > p$, and *optimal* otherwise.

Bibliography

S. Abiteboul, R. Hull, and V. Vianu. *Foundations of Databases.* Addison-Wesley, 1995. 66

J. Alferes and J. Leite, editors. *Proceedings of the Ninth European Conference on Logics in Artificial Intelligence (JELIA'04)*, volume 3229 of *Lecture Notes in Computer Science*, 2004. Springer-Verlag. 185, 190, 199, 204

J. Alferes, L. Pereira, H. Przymusinska, and T. Przymusinski. LUPS: A language for updating logic programs. *Artificial Intelligence*, 138(1-2):87–116, 2002. DOI: 10.1016/S0004-3702(02)00183-2 176

M. Alviano, F. Calimeri, W. Faber, N. Leone, and S. Perri. Unfounded sets and well-founded semantics of answer set programs with aggregates. *Journal of Artificial Intelligence Research*, 42: 487–527, 2011. DOI: 10.1613/jair.3432 89

B. Andres, B. Kaufmann, O. Matheis, and T. Schaub. Unsatisfiability-based optimization in clasp. In A. Dovier and V. Santos Costa, editors, *Technical Communications of the Twenty-eighth International Conference on Logic Programming (ICLP'12)*, volume 17, pages 212–221. Leibniz International Proceedings in Informatics (LIPIcs), 2012. 150

C. Anger, K. Konczak, and T. Linke. noMoRe: A system for non-monotonic reasoning under answer set semantics. In Eiter et al. (2001), pages 406–410. 150

C. Anger, M. Gebser, T. Linke, A. Neumann, and T. Schaub. The nomore++ approach to answer set solving. In G. Sutcliffe and A. Voronkov, editors, *Proceedings of the Twelfth International Conference on Logic for Programming, Artificial Intelligence, and Reasoning (LPAR'05)*, volume 3835 of *Lecture Notes in Artificial Intelligence*, pages 95–109. Springer-Verlag, 2005. DOI: 10.1007/11591191 150

C. Anger, M. Gebser, and T. Schaub. Approaching the core of unfounded sets. In Dix and Hunter (2006), pages 58–66. 110

G. Audemard and L. Simon. Predicting learnt clauses quality in modern SAT solvers. In C. Boutilier, editor, *Proceedings of the Twenty-first International Joint Conference on Artificial Intelligence (IJ-CAI'09)*, pages 399–404. AAAI/MIT Press, 2009. 110, 151

M. Balduccini. Representing constraint satisfaction problems in answer set programming. In W. Faber and J. Lee, editors, *Proceedings of the Second Workshop on Answer Set Programming and Other Computing Paradigms (ASPOCP'09)*, pages 16–30, 2009. 151

M. Balduccini, E. Pontelli, O. El-Khatib, and H. Le. Issues in parallel execution of non-monotonic reasoning systems. *Parallel Computing*, 31(6):608–647, 2005. DOI: 10.1016/j.parco.2005.03.004 150

C. Baral. *Knowledge Representation, Reasoning and Declarative Problem Solving*. Cambridge University Press, 2003. DOI: 10.1017/CBO9780511543357 9, 49

C. Baral and M. Gelfond. Logic programming and knowledge representation. *Journal of Logic Programming*, 12:1–80, 1994. DOI: 10.1016/0743-1066(94)90025-6 49

C. Baral, G. Greco, N. Leone, and G. Terracina, editors. *Proceedings of the Eighth International Conference on Logic Programming and Nonmonotonic Reasoning (LPNMR'05)*, volume 3662 of *Lecture Notes in Artificial Intelligence*, 2005. Springer-Verlag. DOI: 10.1007/11546207 191, 195

C. Baral, G. Brewka, and J. Schlipf, editors. *Proceedings of the Ninth International Conference on Logic Programming and Nonmonotonic Reasoning (LPNMR'07)*, volume 4483 of *Lecture Notes in Artificial Intelligence*, 2007. Springer-Verlag. 185, 188, 192

C. Barrett, R. Sebastiani, S. Seshia, and C. Tinelli. Satisfiability modulo theories. In Biere et al. (2009), chapter 26, pages 825–885. DOI: 10.1007/11513988_4 151

D. Basak, S. Pal, and D. Patranabis. Support vector regression. *Neural Information Processing — Letters and Reviews*, 11(10):203–224, 2007. 150

S. Baselice, P. Bonatti, and M. Gelfond. Towards an integration of answer set and constraint solving. In M. Gabbrielli and G. Gupta, editors, *Proceedings of the Twenty-first International Conference on Logic Programming (ICLP'05)*, volume 3668 of *Lecture Notes in Computer Science*, pages 52–66. Springer-Verlag, 2005. 151

R. Bayardo and R. Schrag. Using CSP look-back techniques to solve real-world SAT instances. In *Proceedings of the Fourteenth National Conference on Artificial Intelligence (AAAI'97)*, pages 203–208. AAAI/MIT Press, 1997. 110

P. Beame and T. Pitassi. Propositional proof complexity: Past, present, and future. *Bulletin of the European Association for Theoretical Computer Science*, 65:66–89, 1998. DOI: 10.1145/602382.602406 90, 110

P. Beame, H. Kautz, and A. Sabharwal. Towards understanding and harnessing the potential of clause learning. *Journal of Artificial Intelligence Research*, 22:319–351, 2004. DOI: 10.1613/jair.1410 90, 110

P. Besnard. *An Introduction to Default Logic*. Symbolic Computation — Artifical Intelligence. Springer-Verlag, 1989. 10

N. Bidoit and C. Froidevaux. Minimalism subsumes default logic and circumscription in stratified logic programming. In *Proceedings of the Second Annual Symposium on Logic in Computer Science (LICS'87)*, pages 89–97. IEEE Computer Society, 1987. 10

A. Biere. PicoSAT essentials. *Journal on Satisfiability, Boolean Modeling and Computation*, 4:75–97, 2008. 151

A. Biere. Lingeling, Plingeling, PicoSAT and PrecoSAT at SAT race 2010. Technical Report 10/1, Institute for Formal Models and Verification. Johannes Kepler University, 2010. 151

A. Biere, M. Heule, H. van Maaren, and T. Walsh, editors. *Handbook of Satisfiability*, volume 185 of *Frontiers in Artificial Intelligence and Applications*. IOS Press, 2009. 9, 90, 110, 184, 187, 198, 199, 202, 203

R. Bihlmeyer, W. Faber, G. Ielpa, V. Lio, and G. Pfeifer. DLV — user manual. URL http://www.dlvsystem.com/man. 149

G. Boenn, M. Brain, M. de Vos, and J. Fitch. Automatic composition of melodic and harmonic music by answer set programming. In Garcia de la Banda and Pontelli (2008), pages 160–174. DOI: 10.1007/978-3-540-89982-2_21 175

A. Bösel, T. Linke, and T. Schaub. Profiling answer set programming: The visualization component of the noMoRe system. In Alferes and Leite (2004), pages 702–705. DOI: 10.1007/978-3-540-30227-8_61 176

M. Brain and M. de Vos. Debugging logic programs under the answer set semantics. In M. de Vos and A. Provetti, editors, *Proceedings of the Third International Workshop on Answer Set Programming (ASP'05)*, volume 142, pages 141–152. CEUR Workshop Proceedings (CEUR-WS.org), 2005. URL http://ceur-ws.org/Vol-142. 177

M. Brain and M. de Vos. The significance of memory costs in answer set solver implementation. *Journal of Logic and Computation*, 19(4):615–641, 2009. DOI: 10.1093/logcom/exn038 150

M. Brain, T. Crick, M. de Vos, and J. Fitch. TOAST: Applying answer set programming to super-optimisation. In Etalle and Truszczyński (2006), pages 270–284. DOI: 10.1007/11799573_21 175

M. Brain, M. Gebser, J. Pührer, T. Schaub, H. Tompits, and S. Woltran. Debugging ASP programs by means of ASP. In Baral et al. (2007), pages 31–43. DOI: 10.1007/978-3-540-72200-7_5 177

M. Brain, O. Cliffe, and M. de Vos. A pragmatic programmer's guide to answer set programming. In de Vos and Schaub (2009), pages 49–63. URL http://ceur-ws.org/Vol-546. 49

G. Brewka. Logic programming with ordered disjunction. In *Proceedings of the National Conference on Artificial Intelligence (AAAI)*, pages 100–105. AAAI Press, 2002. 176

G. Brewka and T. Eiter. Preferred answer sets for extended logic programs. *Artificial Intelligence*, 109(1-2):297–356, 1999. DOI: 10.1016/S0004-3702(99)00015-6 176

G. Brewka, S. Coradeschi, A. Perini, and P. Traverso, editors. *Proceedings of the Seventeenth European Conference on Artificial Intelligence (ECAI'06)*, 2006. IOS Press. 195, 201

G. Brewka, T. Eiter, and M. Truszczyński. Answer set programming at a glance. *Communications of the ACM*, 54(12):92–103, 2011. DOI: 10.1145/2043174.2043195 9

G. Brewka, T. Eiter, and S. McIlraith, editors. *Proceedings of the Thirteenth International Conference on Principles of Knowledge Representation and Reasoning (KR'12)*, 2012. AAAI Press. 194, 199

P. Cabalar and P. Ferraris. Propositional theories are strongly equivalent to logic programs. *Theory and Practice of Logic Programming*, 7(6):745–759, 2007. DOI: 10.1017/S1471068407003110 33

M. Cadoli and A. Schaerf. Compiling problem specifications into SAT. *Artificial Intelligence*, 162 (1-2):89–120, 2005. DOI: 10.1016/j.artint.2004.01.006 10

F. Calimeri, W. Faber, G. Pfeifer, and N. Leone. Pruning operators for disjunctive logic programming systems. *Fundamenta Informaticae*, 71(2-3):183–214, 2006. 110

F. Calimeri, S. Cozza, G. Ianni, and N. Leone. Computable functions in ASP: Theory and implementation. In Garcia de la Banda and Pontelli (2008), pages 407–424. DOI: 10.1007/978-3-540-89982-2_37 67

F. Calimeri, G. Ianni, F. Ricca, M. Alviano, A. Bria, G. Catalano, S. Cozza, W. Faber, O. Febbraro, N. Leone, M. Manna, A. Martello, C. Panetta, S. Perri, K. Reale, M. Santoro, M. Sirianni, G. Terracina, and P. Veltri. The third answer set programming competition: Preliminary report of the system competition track. In Delgrande and Faber (2011), pages 388–403. DOI: 10.1007/978-3-642-20895-9_46 150

F. Calimeri, W. Faber, M. Gebser, G. Ianni, R. Kaminski, T. Krennwallner, N. Leone, F. Ricca, and T. Schaub. ASP-Core-2: Input language format. Available at https://www.mat.unical.it/aspcomp2013/files/ASP-CORE-2.0.pdf, 2012. 151

X. Chen, J. Jiang, J. Ji, G. Jin, and F. Wang. Integrating NLP with reasoning about actions for autonomous agents communicating with humans. In *Proceedings of the IEEE/WIC/ACM International Conference on Intelligent Agent Technology (IAT'09)*, pages 137–140. IEEE, 2009. DOI: 10.1109/WI-IAT.2009.142 175

X. Chen, J. Ji, J. Jiang, G. Jin, F. Wang, and J. Xie. Developing high-level cognitive functions for service robots. In W. van der Hoek, G. Kaminka, Y. Lespérance, M. Luck, and S. Sen, editors, *Proceedings of the Ninth International Conference on Autonomous Agents and Multiagent Systems (AAMAS'10)*, pages 989–996. IFAAMAS, 2010. 175

K. Clark. Negation as failure. In H. Gallaire and J. Minker, editors, *Logic and Data Bases*, pages 293–322. Plenum Press, 1978. DOI: 10.1007/978-1-4684-3384-5 69, 89

E. Clarke, A. Biere, R. Raimi, and Y. Zhu. Bounded model checking using satisfiability solving. *Formal Methods in System Design*, 19(1):7–34, 2001. DOI: 10.1023/A:1011276507260 177

O. Cliffe, M. de Vos, M. Brain, and J. Padget. ASPVIZ: Declarative visualisation and animation using answer set programming. In Garcia de la Banda and Pontelli (2008), pages 724–728. DOI: 10.1007/978-3-540-89982-2_65 176

S. Cook and R. Reckhow. The relative efficiency of propositional proof systems. *Journal of Symbolic Logic*, 44(1):36–50, 1979. DOI: 10.2307/2273702 90

M. D'Agostino and M. Mondadori. The taming of the cut. classical refutations with analytic cut. *Journal of Logic and Computation*, 4(3):285–319, 1994. DOI: 10.1093/logcom/4.3.285 90

M. D'Agostino, D. Gabbay, R. Hähnle, and J. Posegga, editors. *Handbook of Tableau Methods*. Kluwer Academic Publishers, 1999. 89, 201

A. Dal Palù, A. Dovier, E. Pontelli, and G. Rossi. Answer set programming with constraints using lazy grounding. In Hill and Warren (2009), pages 115–129. DOI: 10.1007/978-3-642-02846-5_14 150

E. Dantsin, T. Eiter, G. Gottlob, and A. Voronkov. Complexity and expressive power of logic programming. In *Proceedings of the Twelfth Annual IEEE Conference on Computational Complexity (CCC'97)*, pages 82–101. IEEE Computer Society Press, 1997. DOI: 10.1109/CCC.1997.612304 33

E. Dantsin, T. Eiter, G. Gottlob, and A. Voronkov. Complexity and expressive power of logic programming. *ACM Computing Surveys*, 33(3):374–425, 2001. DOI: 10.1145/502807.502810 10

A. Darwiche and K. Pipatsrisawat. Complete algorithms. In Biere et al. (2009), chapter 3, pages 99–130. 110

M. Davis and H. Putnam. A computing procedure for quantification theory. *Journal of the ACM*, 7: 201–215, 1960. DOI: 10.1145/321033.321034 32, 110

M. Davis, G. Logemann, and D. Loveland. A machine program for theorem-proving. *Communications of the ACM*, 5:394–397, 1962. DOI: 10.1145/368273.368557 32, 110

M. de Vos and T. Schaub, editors. *Proceedings of the First Workshop on Software Engineering for Answer Set Programming (SEA'07)*, volume 281, 2007. CEUR Workshop Proceedings. URL http://ceur-ws.org/Vol-281. 177, 196

188 BIBLIOGRAPHY

M. de Vos and T. Schaub, editors. *Proceedings of the Second Workshop on Software Engineering for Answer Set Programming (SEA'09)*, volume 546, 2009. CEUR Workshop Proceedings. URL http://ceur-ws.org/Vol-546. 177, 185

R. Dechter. *Constraint Processing*. Morgan Kaufmann Publishers, 2003. 90, 151

J. Delgrande and W. Faber, editors. *Proceedings of the Eleventh International Conference on Logic Programming and Nonmonotonic Reasoning (LPNMR'11)*, volume 6645 of *Lecture Notes in Artificial Intelligence*, 2011. Springer-Verlag. DOI: 10.1007/978-3-642-20895-9 186, 191, 193, 194, 201

J. Delgrande, T. Schaub, and H. Tompits. A framework for compiling preferences in logic programs. *Theory and Practice of Logic Programming*, 3(2):129–187, 2003. DOI: 10.1017/S1471068402001539 176

J. Delgrande, T. Schaub, H. Tompits, and K. Wang. A classification and survey of preference handling approaches in nonmonotonic reasoning. *Computational Intelligence*, 20(2):308–334, 2004. DOI: 10.1111/j.0824-7935.2004.00240.x 33, 176

J. Delgrande, T. Schaub, and H. Tompits. A preference-based framework for updating logic programs. In Baral et al. (2007), pages 71–83. DOI: 10.1007/978-3-540-72200-7_8 176

J. Delgrande, T. Schaub, H. Tompits, and S. Woltran. A model-theoretic approach to belief change in answer set programming. *ACM Transactions on Computational Logic*, 2012. To appear. 176

M. Denecker and E. Ternovska. A logic of nonmonotone inductive definitions. *ACM Transactions on Computational Logic*, 9(2):14:1–14:52, 2008. DOI: 10.1145/1342991.1342998 10

M. Denecker, J. Vennekens, S. Bond, M. Gebser, and M. Truszczyński. The second answer set programming competition. In Erdem et al. (2009), pages 637–654. DOI: 10.1007/978-3-642-04238-6_75 150

N. Dershowitz, Z. Hanna, and A. Nadel. Towards a better understanding of the functionality of a conflict-driven SAT solver. In Marques-Silva and Sakallah (2007), pages 287–293. DOI: 10.1007/978-3-540-72788-0_27 110

Y. Dimopoulos, B. Nebel, and J. Köhler. Encoding planning problems in nonmonotonic logic programs. In S. Steel and R. Alami, editors, *Proceedings of the Fourth European Conference on Planning*, volume 1348 of *Lecture Notes in Artificial Intelligence*, pages 169–181. Springer-Verlag, 1997. URL citeseer.nj.nec.com/dimopoulos97encoding.html. 175

J. Dix and A. Hunter, editors. *Proceedings of the Eleventh International Workshop on Nonmonotonic Reasoning (NMR'06)*, number IFI-06-04 in Technical Report Series, 2006. Clausthal University of Technology, Institute for Informatics. 183, 204

J. Dix, U. Furbach, and I. Niemelä. Nonmonotonic reasoning: Towards efficient calculi and implementations. In Robinson and Voronkov (2001), pages 1241–1354. 90

C. Drescher, M. Gebser, T. Grote, B. Kaufmann, A. König, M. Ostrowski, and T. Schaub. Conflict-driven disjunctive answer set solving. In G. Brewka and J. Lang, editors, *Proceedings of the Eleventh International Conference on Principles of Knowledge Representation and Reasoning (KR'08)*, pages 422–432. AAAI Press, 2008. 90, 110, 149, 150

N. Eén and A. Biere. Effective preprocessing in SAT through variable and clause elimination. In F. Bacchus and T. Walsh, editors, *Proceedings of the Eighth International Conference on Theory and Applications of Satisfiability Testing (SAT'05)*, volume 3569 of *Lecture Notes in Computer Science*, pages 61–75. Springer-Verlag, 2005. 151

N. Eén and N. Sörensson. Temporal induction by incremental SAT solving. *Electronic Notes in Theoretical Computer Science*, 89(4), 2003. DOI: 10.1016/S1571-0661(05)82542-3 151, 177

N. Eén and N. Sörensson. An extensible SAT-solver. In E. Giunchiglia and A. Tacchella, editors, *Proceedings of the Sixth International Conference on Theory and Applications of Satisfiability Testing (SAT'03)*, volume 2919 of *Lecture Notes in Computer Science*, pages 502–518. Springer-Verlag, 2004. 110, 151

T. Eiter and M. Fink. Uniform equivalence of logic programs under the stable model semantics. In C. Palamidessi, editor, *Proceedings of the Nineteenth International Conference on Logic Programming (ICLP'03)*, volume 2916 of *Lecture Notes in Computer Science*, pages 224–238. Springer-Verlag, 2003. 176

T. Eiter and A. Polleres. Towards automated integration of guess and check programs in answer set programming: a meta-interpreter and applications. *Theory and Practice of Logic Programming*, 6 (1-2):23–60, 2006. DOI: 10.1017/S1471068405002577 67

T. Eiter, W. Faber, and M. Truszczyński, editors. *Proceedings of the Sixth International Conference on Logic Programming and Nonmonotonic Reasoning (LPNMR'01)*, volume 2173 of *Lecture Notes in Computer Science*, 2001. Springer-Verlag. 183, 196, 204

T. Eiter, M. Fink, G. Sabbatini, and H. Tompits. On properties of update sequences based on causal rejection. *Theory and Practice of Logic Programming*, 2(6):711–767, 2002. DOI: 10.1017/S1471068401001247 176

T. Eiter, W. Faber, N. Leone, and G. Pfeifer. Computing preferred answer sets by meta-interpretation in answer set programming. *Theory and Practice of Logic Programming*, 3(4-5):463–498, 2003. DOI: 10.1017/S1471068403001753 67, 178

T. Eiter, G. Ianni, R. Schindlauer, and H. Tompits. DLVHEX: A prover for semantic-web reasoning under the answer-set semantics. In *Proceedings of the International Conference on Web Intelligence (WI'06)*, pages 1073–1074. IEEE Computer Society, 2006. DOI: 10.1109/WI.2006.64 176

T. Eiter, G. Ianni, T. Lukasiewicz, R. Schindlauer, and H. Tompits. Combining answer set programming with description logics for the semantic web. *Artificial Intelligence*, 172(12-13):1495–1539, 2008. DOI: 10.1016/j.artint.2008.04.002 176

T. Eiter, G. Ianni, and T. Krennwallner. Answer Set Programming: A Primer. In S. Tessaris, E. Franconi, T. Eiter, C. Gutierrez, S. Handschuh, M. Rousset, and R. Schmidt, editors, *Fifth International Reasoning Web Summer School (RW'09)*, volume 5689 of *Lecture Notes in Computer Science*, pages 40–110. Springer-Verlag, 2009. URL http://www.kr.tuwien.ac.at/staff/tkren/pub/2009/rw2009-asp.pdf. 9, 32, 49

E. Ellguth, M. Gebser, M. Gusowski, R. Kaminski, B. Kaufmann, S. Liske, T. Schaub, L. Schneidenbach, and B. Schnor. A simple distributed conflict-driven answer set solver. In Erdem et al. (2009), pages 490–495. DOI: 10.1007/978-3-642-04238-6_47 150

H. Enderton. *A Mathematical Introduction to Logic*. Academic Press, 1972. 31

E. Erdem and F. Türe. Efficient haplotype inference with answer set programming. In Fox and Gomes (2008), pages 436–441. 175

E. Erdem, F. Lin, and T. Schaub, editors. *Proceedings of the Tenth International Conference on Logic Programming and Nonmonotonic Reasoning (LPNMR'09)*, volume 5753 of *Lecture Notes in Artificial Intelligence*, 2009. Springer-Verlag. DOI: 10.1007/978-3-642-04238-6 188, 190, 192, 193, 196, 197

E. Erdem, K. Haspalamutgil, C. Palaz, V. Patoglu, and T. Uras. Combining high-level causal reasoning with low-level geometric reasoning and motion planning for robotic manipulation. In *Proceedings of the IEEE International Conference on Robotics and Automation (ICRA'11)*, pages 4575–4581. IEEE, 2011. DOI: 10.1109/ICRA.2011.5980160 175

S. Etalle and M. Truszczyński, editors. *Proceedings of the Twenty-second International Conference on Logic Programming (ICLP'06)*, volume 4079 of *Lecture Notes in Computer Science*, 2006. Springer-Verlag. 185, 191, 202

W. Faber, N. Leone, C. Mateis, , and G. Pfeifer. Using database optimization techniques for non-monotonic reasoning. In *Proceedings of the Seventh International Workshop on Deductive Databases and Logic Programming (DDLP'99)*, pages 135–139, 1999. 178

W. Faber, N. Leone, and G. Pfeifer. Recursive aggregates in disjunctive logic programs: Semantics and complexity. In Alferes and Leite (2004), pages 200–212. DOI: 10.1007/978-3-540-30227-8_19 31

F. Fages. Consistency of Clark's completion and the existence of stable models. *Journal of Methods of Logic in Computer Science*, 1:51–60, 1994. 89

O. Febbraro, K. Reale, and F. Ricca. ASPIDE: Integrated development environment for answer set programming. In Delgrande and Faber (2011), pages 317–330. DOI: 10.1007/978-3-642-20895-9_37 176

P. Ferraris. Answer sets for propositional theories. In Baral et al. (2005), pages 119–131. DOI: 10.1007/11546207_10 32, 33

P. Ferraris and V. Lifschitz. Mathematical foundations of answer set programming. In S. Artëmov, H. Barringer, A. d'Avila Garcez, L. Lamb, and J. Woods, editors, *We Will Show Them! Essays in Honour of Dov Gabbay*, volume 1, pages 615–664. College Publications, 2005. 9

P. Ferraris, J. Lee, and V. Lifschitz. A new perspective on stable models. In Veloso (2007), pages 372–379. 177

M. Fitting. A Kripke-Kleene semantics for logic programs. *Journal of Logic Programming*, 2(4): 295–312, 1985. DOI: 10.1016/S0743-1066(85)80005-4 89

D. Fox and C. Gomes, editors. *Proceedings of the Twenty-third National Conference on Artificial Intelligence (AAAI'08)*, 2008. AAAI Press. 190, 192

J. Freeman. *Improvements to propositional satisfiability search algorithms*. PhD thesis, University of Pennsylvania, 1995. 151

J. Gallier. *Logic for Computer Science: Foundations of Automated Theorem Proving*. Harper and Row, New York, 1986. 31

E. Gamma, R. Helm, R. Johnson, and J. Vlissides. *Design Patterns: Elements of Reusable Object-Oriented Software*. Addison-Wesley, 1994. 151

M. Garcia de la Banda and E. Pontelli, editors. *Proceedings of the Twenty-fourth International Conference on Logic Programming (ICLP'08)*, volume 5366 of *Lecture Notes in Computer Science*, 2008. Springer-Verlag. 185, 186, 187, 192, 198

M. Gebser and T. Schaub. Loops: Relevant or redundant? In Baral et al. (2005), pages 53–65. DOI: 10.1007/11546207_5 89

M. Gebser and T. Schaub. Tableau calculi for answer set programming. In Etalle and Truszczyński (2006), pages 11–25. DOI: 10.1007/11799573_4 89

M. Gebser and T. Schaub. Characterizing ASP inferences by unit propagation. In E. Giunchiglia, V. Marek, D. Mitchell, and E. Ternovska, editors, *Proceedings of the First International Workshop on Search and Logic: Answer Set Programming and SAT (LaSh'06)*, pages 41–56, 2006b. 90

M. Gebser and T. Schaub. Generic tableaux for answer set programming. In V. Dahl and I. Niemelä, editors, *Proceedings of the Twenty-third International Conference on Logic Programming (ICLP'07)*, volume 4670 of *Lecture Notes in Computer Science*, pages 119–133. Springer-Verlag, 2007. 89

M. Gebser and T. Schaub. Tableau calculi for logic programs under answer set semantics. *ACM Transactions on Computational Logic*, 2012. To appear. 89

M. Gebser, R. Kaminski, B. Kaufmann, M. Ostrowski, T. Schaub, and S. Thiele. A user's guide to gringo, clasp, clingo, and iclingo. URL http://potassco.sourceforge.net. 32, 49, 149

M. Gebser, B. Kaufmann, A. Neumann, and T. Schaub. Conflict-driven answer set solving. In Veloso (2007), pages 386–392. DOI: 10.1016/j.artint.2012.04.001 90, 110, 149

M. Gebser, B. Kaufmann, A. Neumann, and T. Schaub. Conflict-driven answer set enumeration. In Baral et al. (2007), pages 136–148. DOI: 10.1007/978-3-540-72200-7_13 111, 149

M. Gebser, L. Liu, G. Namasivayam, A. Neumann, T. Schaub, and M. Truszczyński. The first answer set programming system competition. In Baral et al. (2007), pages 3–17. DOI: 10.1007/978-3-540-72200-7_3 150

M. Gebser, T. Schaub, and S. Thiele. Gringo: A new grounder for answer set programming. In Baral et al. (2007), pages 266–271. DOI: 10.1007/978-3-540-72200-7_24 67, 149

M. Gebser, T. Janhunen, M. Ostrowski, T. Schaub, and S. Thiele. A versatile intermediate language for answer set programming: Syntax proposal. Unpublished draft, 2008a. URL http://www.cs.uni-potsdam.de/wv/pdfformat/gejaosscth08a.pdf. 149

M. Gebser, R. Kaminski, B. Kaufmann, M. Ostrowski, T. Schaub, and S. Thiele. Engineering an incremental ASP solver. In Garcia de la Banda and Pontelli (2008), pages 190–205. DOI: 10.1007/978-3-540-89982-2_23 151, 176

M. Gebser, B. Kaufmann, A. Neumann, and T. Schaub. Advanced preprocessing for answer set solving. In M. Ghallab, C. Spyropoulos, N. Fakotakis, and N. Avouris, editors, *Proceedings of the Eighteenth European Conference on Artificial Intelligence (ECAI'08)*, pages 15–19. IOS Press, 2008c. 149

M. Gebser, J. Pührer, T. Schaub, and H. Tompits. A meta-programming technique for debugging answer-set programs. In Fox and Gomes (2008), pages 448–453. 67, 177

M. Gebser, R. Kaminski, B. Kaufmann, and T. Schaub. On the implementation of weight constraint rules in conflict-driven ASP solvers. In Hill and Warren (2009), pages 250–264. DOI: 10.1007/978-3-642-02846-5_23 90, 110, 149

M. Gebser, R. Kaminski, M. Ostrowski, T. Schaub, and S. Thiele. On the input language of ASP grounder gringo. In Erdem et al. (2009), pages 502–508. DOI: 10.1007/978-3-642-04238-6_49 149

M. Gebser, B. Kaufmann, and T. Schaub. Solution enumeration for projected Boolean search problems. In W. van Hoeve and J. Hooker, editors, *Proceedings of the Sixth International Conference on Integration of AI and OR Techniques in Constraint Programming for Combinatorial Optimization Problems (CPAIOR'09)*, volume 5547 of *Lecture Notes in Computer Science*, pages 71–86. Springer-Verlag, 2009c. 111, 149

M. Gebser, B. Kaufmann, and T. Schaub. The conflict-driven answer set solver clasp: Progress report. In Erdem et al. (2009), pages 509–514. DOI: 10.1007/978-3-642-04238-6_50 149

M. Gebser, M. Ostrowski, and T. Schaub. Constraint answer set solving. In Hill and Warren (2009), pages 235–249. DOI: 10.1007/978-3-642-02846-5_22 151

M. Gebser, T. Grote, and T. Schaub. Coala: A compiler from action languages to ASP. In T. Janhunen and I. Niemelä, editors, *Proceedings of the Twelfth European Conference on Logics in Artificial Intelligence (JELIA'10)*, volume 6341 of *Lecture Notes in Artificial Intelligence*, pages 360–364. Springer-Verlag, 2010a. 151

M. Gebser, C. Guziolowski, M. Ivanchev, T. Schaub, A. Siegel, S. Thiele, and P. Veber. Repair and prediction (under inconsistency) in large biological networks with answer set programming. In F. Lin and U. Sattler, editors, *Proceedings of the Twelfth International Conference on Principles of Knowledge Representation and Reasoning (KR'10)*, pages 497–507. AAAI Press, 2010b. 175

M. Gebser, T. Grote, R. Kaminski, and T. Schaub. Reactive answer set programming. In Delgrande and Faber (2011), pages 54–66. DOI: 10.1007/978-3-642-20895-9_7 151, 176, 177

M. Gebser, R. Kaminski, B. Kaufmann, and T. Schaub. Challenges in answer set solving. In M. Balduccini and T. Son, editors, *Logic Programming, Knowledge Representation, and Nonmonotonic Reasoning: Essays Dedicated to Michael Gelfond on the Occasion of His 65th Birthday*, volume 6565 of *Lecture Notes in Computer Science*, pages 74–90. Springer-Verlag, 2011b. 178

M. Gebser, R. Kaminski, B. Kaufmann, and T. Schaub. Multi-criteria optimization in answer set programming. In J. Gallagher and M. Gelfond, editors, *Technical Communications of the Twenty-seventh International Conference on Logic Programming (ICLP'11)*, volume 11, pages 1–10. Leibniz International Proceedings in Informatics (LIPIcs), 2011c. 149

M. Gebser, R. Kaminski, B. Kaufmann, T. Schaub, M. Schneider, and S. Ziller. A portfolio solver for answer set programming: Preliminary report. In Delgrande and Faber (2011), pages 352–357. DOI: 10.1007/978-3-642-20895-9_40 150

M. Gebser, R. Kaminski, B. Kaufmann, T. Schaub, and B. Schnor. Cluster-based ASP solving with claspar. In Delgrande and Faber (2011), pages 364–369. DOI: 10.1007/978-3-642-20895-9_42 150

M. Gebser, R. Kaminski, M. Knecht, and T. Schaub. plasp: A prototype for PDDL-based planning in ASP. In Delgrande and Faber (2011), pages 358–363. DOI: 10.1007/978-3-642-20895-9_41 151

M. Gebser, R. Kaminski, A. König, and T. Schaub. Advances in gringo series 3. In Delgrande and Faber (2011), pages 345–351. DOI: 10.1007/978-3-642-20895-9_39 149

M. Gebser, R. Kaminski, and T. Schaub. Complex optimization in answer set programming. *Theory and Practice of Logic Programming*, 11(4-5):821–839, 2011h. DOI: 10.1017/S1471068411000329 33, 67, 176, 178

M. Gebser, J. Lee, and Y. Lierler. On elementary loops of logic programs. *Theory and Practice of Logic Programming*, 11(6):953–988, 2011i. DOI: 10.1017/S1471068411000019 89

M. Gebser, O. Sabuncu, and T. Schaub. An incremental answer set programming based system for finite model computation. *AI Communications*, 24(2):195–212, 2011j. DOI: 10.3233/AIC-2011-0496 151

M. Gebser, T. Schaub, S. Thiele, and P. Veber. Detecting inconsistencies in large biological networks with answer set programming. *Theory and Practice of Logic Programming*, 11(2-3):323–360, 2011k. DOI: 10.1017/S1471068410000554 175

M. Gebser, T. Grote, R. Kaminski, P. Obermeier, O. Sabuncu, and T. Schaub. Stream reasoning with answer set programming: Preliminary report. In Brewka et al. (2012), pages 613–617. 151, 177

M. Gebser, B. Kaufmann, and T. Schaub. Conflict-driven answer set solving: From theory to practice. *Artificial Intelligence*, 187:52–89, 2012b. DOI: 10.1016/j.artint.2012.04.001 90, 149

M. Gebser, B. Kaufmann, and T. Schaub. Multi-threaded ASP solving with clasp. *Theory and Practice of Logic Programming*, 12(4-5):525–545, 2012c. DOI: 10.1017/S1471068412000166 149

gecode. URL http://www.gecode.org. 151

M. Gelfond. On stratified autoepistemic theories. In K. Forbus and H. Shrobe, editors, *Proceedings of the Sixth National Conference on Artificial Intelligence (AAAI'87)*, pages 207–211. Morgan Kaufmann Publishers, 1987. 10

M. Gelfond. Representing knowledge in A-prolog. In A. Kakas and F. Sadri, editors, *Computational Logic: Logic Programming and Beyond, Essays in Honour of Robert Kowalski*, volume 2408 of *Lecture Notes in Computer Science*, pages 413–451. Springer-Verlag, 2002. 9

M. Gelfond. Answer sets. In Lifschitz et al. (2008), chapter 7, pages 285–316. DOI: 10.1016/S1574-6526(07)03007-6 9

M. Gelfond and N. Leone. Logic programming and knowledge representation — the A-Prolog perspective. *Artificial Intelligence*, 138(1-2):3–38, 2002. DOI: 10.1016/S0004-3702(02)00206-0 9

M. Gelfond and V. Lifschitz. The stable model semantics for logic programming. In R. Kowalski and K. Bowen, editors, *Proceedings of the Fifth International Conference and Symposium of Logic Programming (ICLP'88)*, pages 1070–1080. MIT Press, 1988. 10, 13, 31

M. Gelfond and V. Lifschitz. Logic programs with classical negation. In D. Warren and P. Szeredi, editors, *Proceedings of the Seventh International Conference on Logic Programming (ICLP'90)*, pages 579–597. MIT Press, 1990. 31

M. Gelfond and V. Lifschitz. Classical negation in logic programs and disjunctive databases. *New Generation Computing*, 9:365–385, 1991. DOI: 10.1007/BF03037169 10, 31, 33

M. Ginsberg, editor. *Readings in Nonmonotonic Reasoning*. Morgan Kaufmann Publishers, 1987. 10

E. Giunchiglia, Y. Lierler, and M. Maratea. Answer set programming based on propositional satisfiability. *Journal of Automated Reasoning*, 36(4):345–377, 2006. DOI: 10.1007/s10817-006-9033-2 110, 150

E. Goldberg and Y. Novikov. BerkMin: A fast and robust SAT solver. In *Proceedings of the Fifth Conference on Design, Automation and Test in Europe (DATE'02)*, pages 142–149. IEEE Computer Society Press, 2002. DOI: 10.1016/j.dam.2006.10.007 151

G. Grasso, S. Iiritano, N. Leone, V. Lio, F. Ricca, and F. Scalise. An ASP-based system for team-building in the Gioia-Tauro seaport. In M. Carro and R. Peña, editors, *Proceedings of the Twelfth International Symposium on Practical Aspects of Declarative Languages (PADL'10)*, volume 5937 of *Lecture Notes in Computer Science*, pages 40–42. Springer-Verlag, 2010. DOI: 10.1007/978-3-642-11503-5 175

J. Gressmann, T. Janhunen, R. Mercer, T. Schaub, S. Thiele, and R. Tichy. Platypus: A platform for distributed answer set solving. In Baral et al. (2005), pages 227–239. DOI: 10.1007/11546207_18 150

J. Gressmann, T. Janhunen, R. Mercer, T. Schaub, S. Thiele, and R. Tichy. On probing and multi-threading in platypus. In Brewka et al. (2006), pages 392–396. 150

W. Gropp, E. Lusk, and R. Thakur. *Using MPI-2: Advanced Features of the Message-Passing Interface*. MIT Press, 1999. 150

K. Gödel. Zum intuitionistischen Aussagenkalkül. In *Anzeiger der Akademie der Wissenschaften in Wien*, page 65–66. 1932. 10

R. Hähnle. Tableaux and related methods. In Robinson and Voronkov (2001), pages 100–178. DOI: 10.1016/B978-044450813-3/50005-9 89

K. Heljanko and I. Niemelä. Bounded LTL model checking with stable models. *Theory and Practice of Logic Programming*, 3(4-5):519–550, 2003. 175

A. Heyting. Die formalen Regeln der intuitionistischen Logik. In *Sitzungsberichte der Preussischen Akademie der Wissenschaften*, page 42–56. 1930. Reprint in Logik-Texte: Kommentierte Auswahl zur Geschichte der Modernen Logik, Akademie-Verlag, 1986. 10

P. Hill and D. Warren, editors. *Proceedings of the Twenty-fifth International Conference on Logic Programming (ICLP'09)*, volume 5649 of *Lecture Notes in Computer Science*, 2009. Springer-Verlag. 187, 192, 193, 198

J. Huang. The effect of restarts on the efficiency of clause learning. In Veloso (2007), pages 2318–2323. 151

K. Inoue, M. Koshimura, and R. Hasegawa. Embedding negation as failure into a model generation theorem prover. In D. Kapur, editor, *Proceedings of the Eleventh International Conference on Automated Deduction*, volume 607 of *Lecture Notes in Artificial Intelligence*, pages 400–415. Springer-Verlag, 1992. 110, 150

H. Ishebabi, P. Mahr, C. Bobda, M. Gebser, and T. Schaub. Answer set vs integer linear programming for automatic synthesis of multiprocessor systems from real-time parallel programs. *Journal of Reconfigurable Computing*, 2009. URL http://www.hindawi.com/journals/ijrc/2009/863630.html. Article ID 863630. DOI: 10.1155/2009/863630 175

T. Janhunen. URL http://www.tcs.hut.fi/Software/asptools. 151

T. Janhunen. On the effect of default negation on the expressiveness of disjunctive rules. In Eiter et al. (2001), pages 93–106. DOI: 10.1007/3-540-45402-0_7 33

T. Janhunen. Some (in)translatability results for normal logic programs and propositional theories. *Journal of Applied Non-Classical Logics*, 16(1-2):35–86, 2006. DOI: 10.3166/jancl.16.35-86 150

T. Janhunen. Intermediate languages of ASP systems and tools. In de Vos and Schaub (2007), pages 12–25. URL http://ceur-ws.org/Vol-281. 149

T. Janhunen, I. Niemelä, D. Seipel, P. Simons, and J. You. Unfolding partiality and disjunctions in stable model semantics. *ACM Transactions on Computational Logic*, 7(1):1–37, 2006. DOI: 10.1145/1119439.1119440 110, 150

T. Janhunen, I. Niemelä, and M. Sevalnev. Computing stable models via reductions to difference logic. In Erdem et al. (2009), pages 142–154. DOI: 10.1007/978-3-642-04238-6_14 150

M. Järvisalo and E. Oikarinen. Extended ASP tableaux and rule redundancy in normal logic programs. *Theory and Practice of Logic Programming*, 8(5-6):691–716, 2008. DOI: 10.1017/S1471068408003578 89

M. Järvisalo, A. Biere, and M. Heule. Blocked clause elimination. In J. Esparza and R. Majumdar, editors, *Proceedings of the Sixteenth International Conference on Tools and Algorithms for the Construction and Analysis of Systems (TACAS'10)*, volume 6015 of *Lecture Notes in Computer Science*, pages 129–144. Springer-Verlag, 2010. DOI: 10.1007/978-3-642-12002-2 151

H. Kautz and B. Selman. Planning as satisfiability. In B. Neumann, editor, *Proceedings of the Tenth European Conference on Artificial Intelligence (ECAI'92)*, pages 359–363. John Wiley & sons, 1992. 177

K. Konczak, T. Linke, and T. Schaub. Graphs and colorings for answer set programming. *Theory and Practice of Logic Programming*, 6(1-2):61–106, 2006. DOI: 10.1017/S1471068405002528 110, 150

R. Kowalski. Algorithm = logic + control. *Communications of the ACM*, 22(7):424–436, 1979. DOI: 10.1145/359131.359136 10

J. Lee. A model-theoretic counterpart of loop formulas. In L. Kaelbling and A. Saffiotti, editors, *Proceedings of the Nineteenth International Joint Conference on Artificial Intelligence (IJCAI'05)*, pages 503–508. Professional Book Center, 2005. 89

C. Lefèvre and P. Nicolas. The first version of a new ASP solver : ASPeRiX. In Erdem et al. (2009), pages 522–527. DOI: 10.1007/978-3-642-04238-6_52 150

J. Leite. *Evolving Knowledge Bases*, volume 81 of *Frontiers of Artificial Intelligence and Applications*. IOS Press, 2003. 176

N. Leone, P. Rullo, and F. Scarcello. Disjunctive stable models: Unfounded sets, fixpoint semantics, and computation. *Information and Computation*, 135(2):69–112, 1997. DOI: 10.1006/inco.1997.2630 89, 110

N. Leone, G. Greco, G. Ianni, V. Lio, G. Terracina, T. Eiter, W. Faber, M. Fink, G. Gottlob, R. Rosati, D. Lembo, M. Lenzerini, M. Ruzzi, E. Kalka, B. Nowicki, and W. Staniszkis. The INFOMIX system for advanced integration of incomplete and inconsistent data. In F. Özcan, editor, *Proceedings of the ACM SIGMOD International Conference on Management of Data (SIGMOD'05)*, pages 915–917. ACM Press, 2005. DOI: 10.1145/1066157 175

N. Leone, G. Pfeifer, W. Faber, T. Eiter, G. Gottlob, S. Perri, and F. Scarcello. The DLV system for knowledge representation and reasoning. *ACM Transactions on Computational Logic*, 7(3): 499–562, 2006. DOI: 10.1145/1149114.1149117 32, 33, 49, 67, 110, 149, 150

C. Li and F. Manyà. MaxSAT. In Biere et al. (2009), chapter 19, pages 613–631. DOI: 10.1007/s10601-010-9097-9 150

Y. Lierler. Abstract answer set solvers with learning. *Theory and Practice of Logic Programming*, 11 (2-3):135–169, 2011. DOI: 10.1017/S1471068410000578 90, 150

Y. Lierler and V. Lifschitz. One more decidable class of finitely ground programs. In Hill and Warren (2009), pages 489–493. DOI: 10.1007/978-3-642-02846-5_40 67

V. Lifschitz. Foundations of logic programming. In G. Brewka, editor, *Principles of Knowledge Representation*, pages 69–127. CSLI Publications, 1996. 9, 89

V. Lifschitz. Answer set planning. In D. de Schreye, editor, *Proceedings of the International Conference on Logic Programming (ICLP'99)*, pages 23–37. MIT Press, 1999. 9, 174

V. Lifschitz. Answer set programming and plan generation. *Artificial Intelligence*, 138(1-2):39–54, 2002. DOI: 10.1016/S0004-3702(02)00186-8 49, 174

V. Lifschitz. Introduction to answer set programming. Unpublished draft, 2004. URL http://www.cs.utexas.edu/users/vl/papers/esslli.ps. DOI: 10.1007/s10472-007-9080-3 9, 31, 32

V. Lifschitz. Twelve definitions of a stable model. In Garcia de la Banda and Pontelli (2008), pages 37–51. DOI: 10.1007/978-3-540-89982-2_8 90

V. Lifschitz and A. Razborov. Why are there so many loop formulas? *ACM Transactions on Computational Logic*, 7(2):261–268, 2006. DOI: 10.1145/1131313.1131316 10

V. Lifschitz and H. Turner. Splitting a logic program. In *Proceedings of the Eleventh International Conference on Logic Programming*, pages 23–37. MIT Press, 1994. 176

V. Lifschitz and T. Woo. Answer sets in general nonmonotonic reasoning (preliminary report). In B. Nebel, C. Rich, and W. Swartout, editors, *Proceedings of the Third International Conference on Principles of Knowledge Representation and Reasoning (KR'92)*, pages 603–614. Morgan Kaufmann Publishers, 1992. 33

V. Lifschitz, D. Pearce, and A. Valverde. Strongly equivalent logic programs. *ACM Transactions on Computational Logic*, 2(4):526–541, 2001. DOI: 10.1145/383779.383783 176

V. Lifschitz, F. van Harmelen, and B. Porter, editors. *Handbook of Knowledge Representation*. Elsevier Science, 2008. 9, 194

F. Lin. *A Study of Nonmonotonic Reasoning*. PhD thesis, Stanford University, 1991. 89

F. Lin and Y. Zhao. ASSAT: computing answer sets of a logic program by SAT solvers. *Artificial Intelligence*, 157(1-2):115–137, 2004. DOI: 10.1016/j.artint.2004.04.004 89, 110, 150

Z. Lin, Y. Zhang, and H. Hernandez. Fast SAT-based answer set solver. In Y. Gil and R. Mooney, editors, *Proceedings of the Twenty-first National Conference on Artificial Intelligence (AAAI'06)*, pages 92–97. AAAI Press, 2006. 110, 150

T. Linke. Graph theoretical characterization and computation of answer sets. In B. Nebel, editor, *Proceedings of the Seventeenth International Joint Conference on Artificial Intelligence (IJCAI'01)*, pages 641–645. Morgan Kaufmann Publishers, 2001. 31, 110

G. Liu, T. Janhunen, and I. Niemelä. Answer set programming via mixed integer programming. In Brewka et al. (2012), pages 32–42. 150

L. Liu and M. Truszczyński. Properties of programs with monotone and convex constraints. In Veloso and Kambhampati (2005), pages 701–706. DOI: 10.1613/jair.2009 150

X. Liu, C. Ramakrishnan, and S. Smolka. Fully local and efficient evaluation of alternating fixed points (extended abstract). In B. Steffen, editor, *TACAS*, volume 1384 of *Lecture Notes in Computer Science*, pages 5–19. Springer-Verlag, 1998. 175

J. Lloyd. *Foundations of Logic Programming*. Symbolic Computation. Springer-Verlag, 1987. DOI: 10.1007/978-3-642-83189-8 9, 10, 31, 32

V. Marek and M. Truszczyński. *Nonmonotonic logic: context-dependent reasoning*. Artifical Intelligence. Springer-Verlag, 1993. 10

V. Marek and M. Truszczyński. Stable models and an alternative logic programming paradigm. In K. Apt, V. Marek, M. Truszczyński, and D. Warren, editors, *The Logic Programming Paradigm: a 25-Year Perspective*, pages 375–398. Springer-Verlag, 1999. 9, 49

M. Mariën, D. Gilis, and M. Denecker. On the relation between ID-logic and answer set programming. In Alferes and Leite (2004), pages 108–120. DOI: 10.1007/978-3-540-30227-8_12 10

J. Marques-Silva and K. Sakallah. GRASP: A search algorithm for propositional satisfiability. *IEEE Transactions on Computers*, 48(5):506–521, 1999. DOI: 10.1109/12.769433 110, 151

J. Marques-Silva and K. Sakallah, editors. *Proceedings of the Tenth International Conference on Theory and Applications of Satisfiability Testing (SAT'07)*, volume 4501 of *Lecture Notes in Computer Science*, 2007. Springer-Verlag. DOI: 10.1007/978-3-540-72788-0 188, 202

J. Marques-Silva, I. Lynce, and S. Malik. Conflict-driven clause learning SAT solvers. In Biere et al. (2009), chapter 4, pages 131–153. DOI: 10.3233/978-1-58603-929-5-131 110

J. McCarthy. Circumscription — a form of nonmonotonic reasoning. *Artificial Intelligence*, 13(1-2): 27–39, 1980. DOI: 10.1016/0004-3702(80)90011-9 89

J. McCarthy. Applications of circumscription to formalizing common-sense knowledge. *Artificial Intelligence*, 28:89–116, 1986. DOI: 10.1016/0004-3702(86)90032-9 89

J. McCarthy. Elaboration tolerance, 1998. URL http://www-formal.stanford.edu/jmc/elaboration.html. 9

V. Mellarkod and M. Gelfond. Integrating answer set reasoning with constraint solving techniques. In J. Garrigue and M. Hermenegildo, editors, *Proceedings of the Ninth International Symposium on Functional and Logic Programming (FLOPS'08)*, volume 4989 of *Lecture Notes in Computer Science*, pages 15–31. Springer-Verlag, 2008. 151

V. Mellarkod, M. Gelfond, and Y. Zhang. Integrating answer set programming and constraint logic programming. *Annals of Mathematics and Artificial Intelligence*, 53(1-4):251–287, 2008. DOI: 10.1007/s10472-009-9116-y 151

J. Minker, editor. *Foundations of Deductive Databases and Logic Programming*. Morgan Kaufmann Publishers, 1988. 10

R. Moore. Semantical considerations on nonmonotonic logics. *Artificial Intelligence*, 25:75–94, 1985. DOI: 10.1016/0004-3702(85)90042-6 10, 89

M. Moskewicz, C. Madigan, Y. Zhao, L. Zhang, and S. Malik. Chaff: Engineering an efficient SAT solver. In *Proceedings of the Thirty-eighth Conference on Design Automation (DAC'01)*, pages 530–535. ACM Press, 2001. DOI: 10.1145/378239.379017 110, 151

I. Niemelä. Logic programs with stable model semantics as a constraint programming paradigm. *Annals of Mathematics and Artificial Intelligence*, 25(3-4):241–273, 1999. DOI: 10.1023/A:1018930122475 9, 10, 49

I. Niemelä and P. Simons. Evaluating an algorithm for default reasoning. In *Working Notes of the IJCAI'95 Workshop on Applications and Implementations of Nonmonotonic Reasoning Systems*, pages 66–72, 1995. 32

I. Niemelä and P. Simons. Smodels: An implementation of the stable model and well-founded semantics for normal logic programs. In J. Dix, U. Furbach, and A. Nerode, editors, *Proceedings of the Fourth International Conference on Logic Programming and Nonmonotonic Reasoning (LP-NMR'97)*, volume 1265 of *Lecture Notes in Artificial Intelligence*, pages 420–429. Springer-Verlag, 1997. DOI: 10.1007/3-540-63255-7 32

R. Nieuwenhuis, A. Oliveras, and C. Tinelli. Solving SAT and SAT modulo theories: From an abstract Davis-Putnam-Logemann-Loveland procedure to DPLL(T). *Journal of the ACM*, 53(6): 937–977, 2006. DOI: 10.1145/1217856.1217859 90, 151

M. Nogueira, M. Balduccini, M. Gelfond, R. Watson, and M. Barry. An A-prolog decision support system for the space shuttle. In I. Ramakrishnan, editor, *Proceedings of the Third International Symposium on Practical Aspects of Declarative Languages (PADL'01)*, volume 1990 of *Lecture Notes in Computer Science*, pages 169–183. Springer-Verlag, 2001. 175

J. Oetsch, J. Pührer, and H. Tompits. Catching the ouroboros: On debugging non-ground answer-set programs. In *Theory and Practice of Logic Programming. Twenty-sixth International Conference on Logic Programming (ICLP'10) Special Issue*, volume 10(4-6), pages 513–529. Cambridge University Press, 2010. DOI: 10.1017/S1471068410000256 67

J. Oetsch, J. Pührer, M. Seidl, H. Tompits, and P. Zwickl. VIDEAS: A development tool for answer-set programs based on model-driven engineering technology. In Delgrande and Faber (2011), pages 382–387. DOI: 10.1007/978-3-642-20895-9_45 176

E. Oikarinen and T. Janhunen. Modular equivalence for normal logic programs. In Brewka et al. (2006), pages 412–416. 176

N. Olivetti. Tableaux for nonmonotonic logics. In D'Agostino et al. (1999), pages 469–528. 90

M. Ostrowski and T. Schaub. ASP modulo CSP: The clingcon system. *Theory and Practice of Logic Programming*, 12(4-5):485–503, 2012. DOI: 10.1017/S1471068412000142 151

D. Pearce. Default logic and constructive logic. In B. Neumann, editor, *Proceedings of the European Conference on Artificial Intelligence*, pages 309–313. John Wiley & sons, 1994. 175

D. Pearce. A new logical characterisation of stable models and answer sets. In J. Dix, L. Pereira, and T. Przymusinski, editors, *Proceedings of the Sixth Workshop on Non-Monotonic Extensions of Logic Programming (NMELP'96)*, volume 1216 of *Lecture Notes in Computer Science*, pages 57–70. Springer-Verlag, 1996. 10, 175

D. Pearce. Equilibrium logic. *Annals of Mathematics and Artificial Intelligence*, 47(1-2):3–41, 2006. DOI: 10.1007/s10472-006-9028-z 10, 175, 177

D. Pearce, I. de Guzmán, and A. Valverde. A tableau calculus for equilibrium entailment. In R. Dyckhoff, editor, *Proceedings of the Ninth International Conference on Automated Reasoning with Analytic Tableaux and Related Methods (TABLEAUX'00)*, volume 1847 of *Lecture Notes in Computer Science*, pages 352–367. Springer-Verlag, 2000. DOI: 10.1007/10722086 90

D. Pearce, H. Tompits, and S. Woltran. Characterising equilibrium logic and nested logic programs: Reductions and complexity. *Theory and Practice of Logic Programming*, 9(5):565–616, 2009. DOI: 10.1017/S147106840999010X 89

S. Perri, F. Scarcello, G. Catalano, and N. Leone. Enhancing DLV instantiator by back-jumping techniques. *Annals of Mathematics and Artificial Intelligence*, 51(2-4):195–228, 2007. DOI: 10.1007/s10472-008-9090-9 67

K. Pipatsrisawat and A. Darwiche. A lightweight component caching scheme for satisfiability solvers. In Marques-Silva and Sakallah (2007), pages 294–299. DOI: 10.1007/978-3-540-72788-0_28 151

K. Pipatsrisawat and A. Darwiche. On the power of clause-learning SAT solvers as resolution engines. *Artificial Intelligence*, 175(2):512–525, 2011. DOI: 10.1016/j.artint.2010.10.002 90, 110

E. Pontelli and T. Son. Justifications for logic programs under answer set semantics. In Etalle and Truszczyński (2006). DOI: 10.1017/S1471068408003633 177

E. Pontelli, M. Balduccini, and F. Bermudez. Non-monotonic reasoning on Beowulf platforms. In V. Dahl and P. Wadler, editors, *Proceedings of the Fifth International Symposium on Practical Aspects of Declarative Languages (PADL'03)*, volume 2562 of *Lecture Notes in Artificial Intelligence*, pages 37–57. Springer-Verlag, 2003. DOI: 10.1007/3-540-36388-2 150

potassco. URL http://potassco.sourceforge.net. DOI: 10.1007/s10992-011-9215-1 150

R. Reiter. On closed world data bases. In H. Gallaire and J. Minker, editors, *Logic and Databases*, pages 55–76. Plenum Press, New York, 1978. 10

R. Reiter. A logic for default reasoning. *Artificial Intelligence*, 13(1-2):81–132, 1980. DOI: 10.1016/0004-3702(80)90014-4 10, 31, 89

F. Ricca. A java wrapper for DLV. In M. de Vos and A. Provetti, editors, *Proceedings of the Second International Workshop on Answer Set Programming (ASP'03)*, volume 78. CEUR Workshop Proceedings (CEUR-WS.org), 2003. URL http://ceur-ws.org/Vol-78. 176

F. Ricca, W. Faber, and N. Leone. A backjumping technique for disjunctive logic programming. *AI Communications*, 19(2):155–172, 2006. 150

F. Ricca, L. Gallucci, R. Schindlauer, T. Dell'Armi, G. Grasso, and N. Leone. OntoDLV: An ASP-based system for enterprise ontologies. *Journal of Logic and Computation*, 19(4):643–670, 2009. DOI: 10.1093/logcom/exn042 176

J. Rintanen. Planning and SAT. In Biere et al. (2009), chapter 15, pages 483–504. DOI: 10.1007/978-3-642-15396-9_34 177

J. Rintanen, K. Heljanko, and I. Niemelä. Planning as satisfiability: parallel plans and algorithms for plan search. *Artificial Intelligence*, 170(12-13):1031–1080, 2006. DOI: 10.1016/j.artint.2006.08.002 174

A. Robinson and A. Voronkov, editors. *Handbook of Automated Reasoning*. Elsevier and MIT Press, 2001. 189, 196

F. Rossi, P. van Beek, and T. Walsh, editors. *Handbook of Constraint Programming*. Elsevier Science, 2006. 9, 90, 151

O. Roussel and V. Manquinho. Pseudo-Boolean and cardinality constraints. In Biere et al. (2009), chapter 22, pages 695–733. DOI: 10.3233/978-1-58603-929-5-695 150

L. Ryan. Efficient algorithms for clause-learning SAT solvers. Master's thesis, Simon Fraser University, 2004. 110, 151

C. Sakama and K. Inoue. Updating extended logic programs through abduction. In M. Gelfond, N. Leone, and G. Pfeifer, editors, *Proceedings of the Fifth International Conference on Logic Programming and Nonmonotonic Reasoning (LPNMR'99)*, volume 1730 of *Lecture Notes in Artificial Intelligence*, pages 147–161. Springer-Verlag, 1999. DOI: 10.1007/3-540-46767-X 176

C. Sakama and K. Inoue. Prioritized logic programming and its application to commonsense reasoning. *Artificial Intelligence*, 123(1-2):185–222, 2000. DOI: 10.1016/S0004-3702(00)00054-0 176

T. Schaub and K. Wang. A semantic framework for preference handling in answer set programming. *Theory and Practice of Logic Programming*, 3(4-5):569–607, 2003. DOI: 10.1017/S1471068403001844 176

J. Schlipf. The expressive powers of the logic programming semantics. *Journal of Computer and System Sciences*, 51:64–86, 1995. DOI: 10.1006/jcss.1995.1053 10

L. Schneidenbach, B. Schnor, M. Gebser, R. Kaminski, B. Kaufmann, and T. Schaub. Experiences running a parallel answer set solver on Blue Gene. In M. Ropo, J. Westerholm, and J. Dongarra, editors, *Proceedings of the Sixteenth European PVM/MPI Users' Group Meeting on Recent Advances in Parallel Virtual Machine and Message Passing Interface (PVM/MPI'09)*, volume 5759 of *Lecture Notes in Computer Science*, pages 64–72. Springer-Verlag, 2009. DOI: 10.1007/978-3-642-03770-2 150

C. Schwind. A tableaux-based theorem prover for a decidable subset of default logic. In M. Stickel, editor, *Proceedings of the Tenth International Conference on Automated Deduction (CADE'90)*, volume 449 of *Lecture Notes in Computer Science*, pages 528–542. Springer-Verlag, 1990. DOI: 10.1007/3-540-52885-7 89

P. Simons. *Extending and Implementing the Stable Model Semantics*. Dissertation, Helsinki University of Technology, 2000. 32

P. Simons, I. Niemelä, and T. Soininen. Extending and implementing the stable model semantics. *Artificial Intelligence*, 138(1-2):181–234, 2002. DOI: 10.1016/S0004-3702(02)00187-X 32, 33, 110, 150

J. Slaney and S. Thiébaux. Blocks world revisited. *Artificial Intelligence*, 125(1-2):119–153, 2001. DOI: 10.1016/S0004-3702(00)00079-5 174

M. Slota and J. Leite. On semantic update operators for answer-set programs. In H. Coelho, R. Studer, and M. Wooldridge, editors, *Proceedings of the Nineteenth European Conference on Artificial Intelligence (ECAI'10)*, pages 957–962. IOS Press, 2010. 176

T. Soininen and I. Niemelä. Developing a declarative rule language for applications in product configuration. In G. Gupta, editor, *Proceedings of the First International Workshop on Practical Aspects of Declarative Languages (PADL'99)*, volume 1551 of *Lecture Notes in Computer Science*, pages 305–319. Springer-Verlag, 1999. 175

T. Syrjänen. Lparse 1.0 user's manual. URL `http://www.tcs.hut.fi/Software/smodels/lparse.ps.gz`. 32, 33, 66, 149

T. Syrjänen. Omega-restricted logic programs. In Eiter et al. (2001), pages 267–279. DOI: 10.1007/3-540-45402-0_20 66, 67

T. Syrjänen. Cardinality constraint programs. In Alferes and Leite (2004), pages 187–199. DOI: 10.1007/978-3-540-30227-8_18 33

T. Syrjänen. Debugging inconsistent answer set programs. In Dix and Hunter (2006), pages 77–83. 177

T. Syrjänen. *Logic Programs and Cardinality Constraints: Theory and Practice*. Dissertation, Aalto University, 2009. 66

A. Tarski. A lattice-theoretic fixpoint theorem and its applications. *Pacific Journal of Mathematics*, 5:285–309, 1955. DOI: 10.2140/pjm.1955.5.285 32

G. Terracina, E. De Francesco, C. Panetta, and N. Leone. Enhancing a DLP system for advanced database applications. In D. Calvanese and G. Lausen, editors, *Proceedings of the Second International Conference on Web Reasoning and Rule Systems (RR'08)*, volume 5341 of *Lecture Notes in Computer Science*, pages 119–134. Springer-Verlag, 2008. DOI: 10.1007/978-3-540-88737-9 176

E. Torlak and D. Jackson. Kodkod: a relational model finder. In *Proceedings of the Thirteenth International Conference on Tools and Algorithms for the Construction and Analysis of Systems (TACAS'07)*, pages 632–647. Springer-Verlag, 2007. DOI: 10.1007/978-3-540-71209-1_49 10

M. Truszczynski. Trichotomy and dichotomy results on the complexity of reasoning with disjunctive logic programs. *Theory and Practice of Logic Programming*, 11(6):881–904, 2011. DOI: 10.1017/S1471068410000463 33

J. Ullman. *Principles of Database and Knowledge-Base Systems*. Computer Science Press, 1988. 9, 66

A. Van Gelder. The alternating fixpoint of logic programs with negation. *Journal of Computer and System Science*, 47:185–221, 1993. DOI: 10.1145/73721.73722 89

A. Van Gelder, K. Ross, and J. Schlipf. The well-founded semantics for general logic programs. *Journal of the ACM*, 38(3):620–650, 1991. DOI: 10.1145/116825.116838 89

M. Veloso, editor. *Proceedings of the Twentieth International Joint Conference on Artificial Intelligence (IJCAI'07)*, 2007. AAAI/MIT Press. 191, 192, 196

M. Veloso and S. Kambhampati, editors. *Proceedings of the Twentieth National Conference on Artificial Intelligence (AAAI'05)*, 2005. AAAI Press. 199, 205

J. Ward and J. Schlipf. Answer set programming with clause learning. In V. Lifschitz and I. Niemelä, editors, *Proceedings of the Seventh International Conference on Logic Programming and Nonmonotonic Reasoning (LPNMR'04)*, volume 2923 of *Lecture Notes in Artificial Intelligence*, pages 302–313. Springer-Verlag, 2004. DOI: 10.1007/b94792 150

S. Woltran. Equivalence between extended datalog programs - a brief survey. In O. de Moor, G. Gottlob, T. Furche, and Sellers A, editors, *First International Workshop on Datalog*, volume 6702 of *Lecture Notes in Computer Science*, pages 106–119. Springer-Verlag, 2011. 176

L. Xu, F. Hutter, H. Hoos, and K. Leyton-Brown. SATzilla: Portfolio-based algorithm selection for SAT. *Journal of Artificial Intelligence Research*, 32:565–606, 2008. DOI: 10.1613/jair.2490 150

L. Zhang, C. Madigan, M. Moskewicz, and S. Malik. Efficient conflict driven learning in a Boolean satisfiability solver. In *Proceedings of the International Conference on Computer-Aided Design (IC-CAD'01)*, pages 279–285, 2001. DOI: 10.1109/ICCAD.2001.968634 110

Y. Zhang and N. Foo. A unified framework for representing logic program updates. In Veloso and Kambhampati (2005), pages 707–713. 176

Index

Printed in the United States
by Baker & Taylor Publisher Services